Differential-Algebraic Equations Forum

Differential-Algebraic Equations Forum

The series "Differential-Algebraic Equations Forum" is concerned with analytical, algebraic, control theoretic and numerical aspects of differential algebraic equations (DAEs) as well as their applications in science and engineering. It is aimed to contain survey and mathematically rigorous articles, research monographs and textbooks. Proposals are assigned to an Associate Editor, who recommends publication on the basis of a detailed and careful evaluation by at least two referees. The appraisals will be based on the substance and quality of the exposition.

For further volumes:
http://www.springer.com/series/11221

Sebastian Schöps • Andreas Bartel •
Michael Günther • E. Jan W. ter Maten •
Peter C. Müller
Editors

Progress in Differential-Algebraic Equations

Deskriptor 2013

Springer

Editors
Sebastian Schöps
Technische Universität Darmstadt
Graduate School CE
Darmstadt
Germany

Andreas Bartel
Applied Mathematics and Numerical Analysis
Bergische Universität Wuppertal
Wuppertal
Germany

Michael Günther
Applied Mathematics and Numerical Analysis
Bergische Univeirsität Wuppertal
Wuppertal
Germany

E. Jan W. ter Maten
Department of Mathematics and Computer Science
Eindhoven University of Technology
Eindhoven
The Netherlands

Peter C. Müller
Department of Safety Control Engineering
Bergische Universität Wuppertal
Wuppertal
Germany

ISBN 978-3-662-44925-7 ISBN 978-3-662-44926-4 (eBook)
DOI 10.1007/978-3-662-44926-4
Springer Heidelberg New York Dordrecht London

Library of Congress Control Number: 2014956018

Mathematics Subject Classification (2010): 34-06, 65-06, 93-06

Printed on acid-free paper

Springer is part of Springer Science+Business Media (www.springer.com)

Preface

The 8th Workshop on Coupled Descriptor Systems took place on March 4–7, 2013 in the Castle of Eringerfeld, Geseke, in the neighborhood of Paderborn, Germany. Following the tradition of the seven preceding workshops organized by Prof. Peter C. Müller between 1992 and 2005, the workshop brought together more than 30 mathematicians and engineers from various fields, e.g., mechanics and electromagnetic theory. The participants focused on the theoretical and numerical treatment of coupled and multiphysical 'descriptor' systems of differential-algebraic equations.

This book contains the proceedings of this workshop. It examines the wide range of current research topics in descriptor systems, including mathematical modeling, index analysis, wellposedness of problems, stiffness and different time-scales, cosimulation and splitting methods and convergence analysis. In addition, it also presents applications from the automotive and circuit industries that show that descriptor systems are still challenging problems from the point of view of theory and practice.

This book is organized into three parts with the first covering **Control**. It features a contribution by Peter C. Müller that discusses *Lyapunov Matrix Equations for the Stability Analysis of Linear Time-Invariant Descriptor Systems*. In it, he compares two different generalizations of the classical Lyapunov matrix equation in theory and in the application to mechanical descriptor systems. This part also includes a contribution by Thomas Berger entitled *Zero Dynamics and Stabilization for Linear DAEs* which is devoted to the study of linear differential-algebraic multi-input multi-output systems. In it, the author introduces the concepts of autonomous zero dynamics, transmission zeros, right-invertibility, stabilizability in the behavioral sense and detectability in the behavioral sense and also derives algebraic characterizations. Also in this part, Daniel Labisch and Ulrich Konigorski introduce new necessary and sufficient conditions for causal observability of nonlinear descriptor systems and present a method to design the causal observer. The approach is based on the transformation of the descriptor system into a state space form.

The second part of this book covers **Simulation** and consists of three contributions: the first contribution *Monitoring Singularities While Integrating DAEs* is by Diana Estévez Schwarz and René Lamour. It presents an approach that allows (higher) index determination during runtime by using automatic (or algorithmic) differentiation. This part also includes two contributions by Martin Arnold. His first article entitled *Modular Time Integration of Block-Structured Coupled Systems Without Algebraic Loops* offers an introduction to the mathematical challenges of cooperative simulation (co-simulation). In his second article *Efficient Time Integration of Block-Structured Descriptor Systems*, Arnold and his co-workers discuss recent results, which were also presented within a keynote lecture at Eringerfeld. This second article has been previously published in the *Archive of Mechanical Engineering* LX(2013) 75–94 (DOI: 10.2478/meceng-2013-0005) and it is reproduced in the present proceedings volume with kind permission of the Polish Academy of Sciences. The last contribution in this part is the work of Lennart Jansen and Caren Tischendorf entitled *A Unified (P)DAE Modeling Approach for Flow Networks*. In it, the authors present a modeling framework for the various types of flow networks, e.g., electric circuits but also water and gas supplying networks in the context of differential-algebraic equations.

The third and final part includes two contributions with a strong focus on **Model Order Reduction**. Nicodemus Banagaaya and Wil H. A. Schilders present a new model-order reduction method for DAEs, which reduces differential-algebraic equations while preserving the index of the system. This paper features examples from circuit simulation. In addition, Sara Grundel and co-workers investigate *Model Order Reduction of Differential Algebraic Equations Arising from the Simulation of Gas Transport Networks*. This paper explores the tractability index of the system and proposes tailored model order reduction techniques.

The workshop and this book would not have been possible without the help of many individuals, including the participants and invited speakers as well as such helpful colleagues as Lara Knist, B.A., at the University of Wuppertal, and Ruth Allewelt from Springer. Also we would like to thank the following for their financial support: the Klaus-Körper-Stiftung that is related to the International Association of Applied Mathematics and Mechanics (GAMM), the Excellence Initiative of the German Federal and State Governments, and the Graduate School of Computational Engineering at Technische Universität Darmstadt.

Darmstadt, Germany — Sebastian Schöps
Andreas Bartel
Wuppertal, Germany — Michael Günther
Eindhoven, The Netherlands — Jan ter Maten
Wuppertal, Germany — Peter C. Müller
August 2014

Contents

Part I Control

Part I
Control

Lyapunov Matrix Equations for the Stability Analysis of Linear Time-Invariant Descriptor Systems

Peter C. Müller

Abstract For the stability analysis of linear time-invariant descriptor systems two different generalizations of the classical Lyapunov matrix equation are considered. The first generalization includes the singular matrix related to the time-derivatives of the descriptor variables in an obviously symmetric form; the second one shows at a first sight no symmetry which additionally has to be asked for explicitly. This second approach is well-known for 'admissible' descriptor systems which includes a restriction to systems of index $k = 1$. In this contribution the second approach will be generalized to systems with arbitrary index $k \geq 1$. Both approaches will be compared with each other showing different solvability conditions and different solutions in general. But for the problem of analyzing asymptotic stability the solution behaviors of the two generalized Lyapunov matrix equations coincide. In spite of the different procedures both approaches lead to the same Lyapunov function for the analysis of asymptotic stability of linear time-invariant descriptor systems. The two approaches will be illustrated by the stability analysis of mechanical descriptor systems, i.e. by mechanical systems with holonomic constrains. Although the application of the approaches usually is very costly, they represent suitable tools for the stability analysis of linear time-invariant descriptor systems.

Keywords Linear descriptor systems • Asymptotic stability • Lyapunov functions • Generalized Lyapunov matrix equations • Mechanical descriptor systems

Mathematics Subject Classification (2010) 93B07 · 93B17 · 93D05

P.C. Müller (✉)
Safety Control Engineering, Bergische Universität Wuppertal, Gaußstraße 20, 42119 Wuppertal, Germany
e-mail: mueller@srm.uni-wuppertal.de

S. Schöps et al. (eds.), *Progress in Differential-Algebraic Equations*,
Differential-Algebraic Equations Forum, DOI 10.1007/978-3-662-44926-4_1

1 Introduction

For linear time-invariant state systems

$$\dot{\mathrm{x}}(t) = \mathrm{A}\,\mathrm{x}(t) \tag{1}$$

it is well-known, cf. e.g. [7], that the asymptotic stability of system (1) can be analysed by a quadratic Lyapunov function

$$v = \frac{1}{2}\mathrm{x}^\top \mathrm{P}\,\mathrm{x} \tag{2}$$

asking for a negative time-derivative $\dot{v} = -\frac{1}{2}\mathrm{x}^\top \mathrm{Q}\mathrm{x}$ which leads to the requirements of the Lyapunov matrix equation:

$$\mathrm{A}^\top \mathrm{P} + \mathrm{P}\,\mathrm{A} = -\mathrm{Q}, \quad \mathrm{P} = \mathrm{P}^\top, \quad \mathrm{Q} = \mathrm{Q}^\top. \tag{3}$$

Then the theorem on asymptotic stability is obtained:

> System (1) is asymptotically stable if and only if Eqs. (3) have a positive definite solution $\mathrm{P} > 0$ for each $\mathrm{Q} = \mathrm{Q}^\top \geq 0$ where (A, Q) is completely observable, cf. [7].

Now the question arises how this result can be generalized for linear time-invariant descriptor systems

$$\mathrm{E}\,\dot{\mathrm{x}}(t) = \mathrm{A}\,\mathrm{x}(t) \tag{4}$$

where the standard assumptions

$$\begin{gathered} \mathrm{rk}\,\mathrm{E} < n = \dim\ \mathrm{x}, \\ \det(s\mathrm{E} - \mathrm{A}) \neq 0 \text{ for one } s \text{ at least} \end{gathered} \tag{5}$$

are required. There are two different approaches. The first one is very natural. A Lyapunov function

$$v_1 = \frac{1}{2}\mathrm{x}^\top \mathrm{E}^\top \mathrm{P}_1 \mathrm{E}\,\mathrm{x} \tag{6}$$

with a symmetric matrix

$$\mathrm{P}_1 = \mathrm{P}_1^\top \tag{7}$$

leads in the usual manner to a generalized Lyapunov matrix equation

$$\mathrm{A}^\top \mathrm{P}_1\,\mathrm{E} + \mathrm{E}^\top \mathrm{P}_1\,\mathrm{A} = -\mathrm{Q} \tag{8}$$

where Q is also symmetric: $Q = Q^T$. This equation has been considered in detail with respect to the problem of asymptotic stability in [14], but it has been applied also in the works [5, 6, 8, 11].

The second approach is based on the Lyapunov function

$$v_2 = \frac{1}{2} x^T E^T P_2 \, x \tag{9}$$

where for symmetry

$$E^T P_2 = P_2^T E \tag{10}$$

is required ($P_2 \neq P_2^T$ in general). Then the usual Lyapunov procedure leads to the second generalized Lyapunov matrix equation

$$A^T P_2 + P_2^T A = -Q \tag{11}$$

where again Q is assumed to be symmetric: $Q = Q^T$. This approach has been applied for 'admissible' systems (see Sect. 3.3) which have been discussed in [12, 15].

The aim of this contribution is the stability analysis of system (4) by these two approaches. While the first approach (8) is well investigated [14], the second approach (10, 11) will be represented here. Then the two approaches will be compared, on the one side the general equations (8) and (11), and on the other side the more special equations (8) and (11) in case of asymptotic stability of system (4).

2 General Reflections on the Generalized Lyapunov Matrix Equations

Here, it is dealt with some general properties of the solutions of Eqs. (7, 8) and (10, 11) without looking for asymptotic stability of system (4).

2.1 *Relation Between Solutions of the Lyapunov Matrix Equations*

Between Eqs. (7, 8) and (10, 11) there exists a simple relation

Lemma 1 *If for a (suitable)* $Q = Q^T$ *Eqs. (7, 8) have a solution* P_1 *then Eqs. (10, 11) have a solution*

$$P_2 = P_1 \, E. \tag{12}$$

Proof Insertion of $P_2 = P_1$ E into (10, 11) shows that these equations are satisfied by the validity of (7, 8).

Lemma 1 states if there is a solution of (7, 8) then there is a solution of (10, 11) also. But nothing is said whether the solution is unique or whether multiple solutions P_1 are related to the same variety of multiple solutions P_2 by Eq. (12). To get some feeling on the solution behaviour of these equations two examples are considered.

2.2 Examples

Differentiator

A dynamical transfer system where the output signal y represents the differentiated input signal u is described by the simple descriptor system

$$\begin{aligned} \dot{x}_2 &= x_1 \\ 0 &= x_2 - u \\ y &= x_1 \end{aligned} \qquad \Longrightarrow \qquad \begin{aligned} x_1 &= \dot{u} \\ x_2 &= u \\ y &= \dot{u} \end{aligned}$$

The system matrices are

$$\mathrm{E} = \begin{bmatrix} 0 & 1 \\ 0 & 0 \end{bmatrix}, \quad \mathrm{A} = \begin{bmatrix} 1 & 0 \\ 0 & 1 \end{bmatrix} = \mathrm{I}_2.$$

Equations (7, 8) result in

$$\text{Solvability}: \quad q_{11} = 0; \qquad \text{Solution}: \quad \mathrm{P}_1 = \begin{bmatrix} -q_{12} & -\frac{1}{2}q_{22} \\ -\frac{1}{2}q_{22} & p_{22} \end{bmatrix}.$$

Equations (10, 11) result in:

$$\text{Solvability}: \quad q_{11} = 0; \qquad \text{Solution}: \quad \mathrm{P}_2 = \begin{bmatrix} 0 & \overline{p}_{12} \\ -q_{12} - \overline{p}_{12} & -\frac{1}{2}q_{22} \end{bmatrix}.$$

Checking Lemma 1 leads to

$$\mathrm{P}_1\mathrm{E} = \begin{bmatrix} 0 & -q_{12} \\ 0 & -\frac{1}{2}q_{22} \end{bmatrix} = \mathrm{P}_2 \quad \text{for} \quad \overline{p}_{12} = -q_{12}.$$

By the Lemma one special solution P_2 is obtained only, but in general P_2 shows a one-dimensional variety of solutions with the parameter $\overline{p}_{12}$. Also P_1 shows a one-dimensional variety of solutions with the parameter p_{22} which is cancelled by the application of the Lemma. The solvability conditions coincide.

Double Differentiator

The first example is extended as follows:

$$\begin{aligned} \dot{x}_2 &= x_1 \\ \dot{x}_3 &= x_2 \\ 0 &= x_3 - u \\ y &= x_1 \end{aligned} \qquad \Longrightarrow \qquad \begin{aligned} x_3 &= u \\ x_2 &= \dot{u} \\ x_1 &= \ddot{u} \\ y &= \ddot{u} \end{aligned}$$

Without input and output the system is characterized by

$$u = 0: \quad \mathrm{E} = \begin{bmatrix} 0 & 1 & 0 \\ 0 & 0 & 1 \\ 0 & 0 & 0 \end{bmatrix}, \quad \mathrm{A} = \mathrm{I} = \begin{bmatrix} 1 & & \\ & 1 & \\ & & 1 \end{bmatrix} = \mathrm{I}_3.$$

Then Eq. (7, 8) lead to following results:

$$\text{Solvability}: \quad q_{11} = 0, \quad q_{13} = \frac{1}{2} q_{22};$$

$$\text{Solution}: \quad \mathrm{P}_1 = \begin{bmatrix} -q_{12} & -q_{13} & -q_{23} - p_{22} \\ -q_{13} & p_{22} & -\frac{1}{2} q_{33} \\ -q_{23} - p_{22} & -\frac{1}{2} q_{33} & p_{33} \end{bmatrix}.$$

Analogously Eqs. (10, 11) result in:

$$\text{Solvability}: \quad q_{11} = 0;$$

$$\text{Solution}: \quad \mathrm{P}_1 = \begin{bmatrix} 0 & -q_{12} & -\frac{1}{2} q_{22} \\ 0 & -\frac{1}{2} q_{22} & -q_{23} - \overline{p}_{32} \\ -q_{13} + \frac{1}{2} q_{22} & \overline{p}_{32} & -\frac{1}{2} q_{33} \end{bmatrix}.$$

Checking Lemma 1 the additional solvability condition $q_{13} = \frac{1}{2} q_{22}$ has to be assumed. Then

$$\mathrm{P}_1 \mathrm{E} = \begin{bmatrix} 0 & -q_{12} & -q_{13} \\ 0 & -q_{13} & p_{22} \\ 0 & -q_{23} - p_{22} & -\frac{1}{2} q_{33} \end{bmatrix} = \mathrm{P}_2$$

holds for the special choice $\overline{p}_{32} = -q_{23} - p_{22}$. By the Lemma a special solution of P_2 is obtained with one free parameter p_{22}, but Q is restricted according to the solvability conditions of P_1. But in this example P_1 requires 2 solvability conditions and has a 2-dimensional variety of solutions according to the parameters p_{22}, p_{33}; P_2 requires 1 solvability condition and has a 1-dimensional variety of solutions according to $\overline{p}_{32}$. The solvability conditions for P_1 and P_2 do not coincide.

In general, the solvability conditions for P_1 and P_2 are different (or coincide partly only). For the solvability conditions of P_1 Lemma 1 can be applied to generate a special solution of P_2. In general, the solutions P_1 and P_2 are not determined uniquely, i.e. a variety of solutions may exist depending on a certain number of parameters. (Remark: The examples consist of fast subsystems only (see Sect. 3) such that it is not surprising to look for solvability conditions and nonunique solutions.)

3 Asymptotic Stability

In this section the two approaches, i.e. Eqs. (7, 8) and (10, 11) will be considered with respect to the analysis of asymptotic stability of the descriptor system (4). For this, firstly the problem will be discussed using the Weierstrass canonical form. Afterwards, the problem will be considered in the original form (4) applying the results which are known by the canonical system representation.

3.1 Weierstrass Canonical Form

For the regular descriptor system (4, 5) there exist regular matrices R, S such that

$$\mathrm{S\,E\,R} = \overline{\mathrm{E}} = \begin{bmatrix} \mathrm{I}_1 & \\ & \mathrm{N} \end{bmatrix}, \quad \mathrm{S\,A\,R} = \overline{\mathrm{A}} = \begin{bmatrix} \mathrm{A}_1 & \\ & \mathrm{I}_2 \end{bmatrix} \tag{13}$$

where N is nilpotent with index k:

$$\mathrm{N}^{k-1} \neq 0, \quad \mathrm{N}^k = 0. \tag{14}$$

This classical result is due to Weierstrass and is represented in detail in [2, 3]. Additionally this transformation can be used for a decoupled system representation of Eq. (4):

$$\mathrm{x} = \mathrm{R}\begin{bmatrix} \mathrm{x}_1 \\ \mathrm{x}_2 \end{bmatrix}, \quad \dim(\mathrm{x}_1) = n_1, \quad \dim(\mathrm{x}_2) = n_2, \quad n_1 + n_2 = n$$

$$\dot{\mathrm{x}}_1 = \mathrm{A}_1\,\mathrm{x}_1, \quad \mathrm{N}\,\dot{\mathrm{x}}_2 = \mathrm{x}_2. \tag{15}$$

The first subsystem (x_1) is called the 'slow' subsystem, which is a state system of order n_1, while the second subsystem (x_2) is called 'fast'. The representation (15) allows the well-known stability characterization of the descriptor system (4, 5), cf. [2, 3]:

Lemma 2 *The descriptor system (4, 5) is asymptotically stable if and only if the slow subsystem is asymptotically stable, i.e.* $\mathrm{Re}\,\lambda_i(A_1) < 0, = 1\ldots, n_1$.

3.2 *First Approach*

The first approach of the stability analysis is defined by Eqs. (7, 8). Defining the (right and left) spectral projection matrices

$$P_r = R\begin{bmatrix} I_1 & \\ & 0 \end{bmatrix}R^{-1}, \quad P_l = S^{-1}\begin{bmatrix} I_1 & \\ & 0 \end{bmatrix}S \tag{16}$$

then following result has been proved [14] and represented again in [6]:

Theorem 1 *The regular descriptor system (4, 5) is asymptotically stable if and only if the projected generalized Lyapunov equation*

$$E^\top P_1\,A + A^\top P_1\,E = -P_r^\top Q\,P_r, \quad P_1 = P_l^\top\,P_1\,P_l = P_1^\top \tag{17}$$

has a unique symmetric, positive semidefinite solution $P_1 = P_1^\top \geq 0$ *for every symmetric, positive definite matrix* $Q = Q^\top > 0$.

Remark 1 Although Theorem 1 holds for all $Q = Q^\top > 0$ the explicit solution of Eqs. (17) may be more or less difficult depending on the choice of Q, particularly if a physical meaning of Q and P_1 is expected. Therefore a certain weaking of Theorem 1 could be helpful. E.g. for mechanical descriptor systems, it is difficult to find a suitable $Q = Q^\top > 0$ to satisfy (17) but it is easy to have a positive semidefinite $Q = Q^\top \geq 0$. In this case the requirement $Q = Q^\top > 0$ can be replaced by the requirements $Q = Q^\top \geq 0$ and R-observability of (E, A; Q).

Remark 2 A certain disadvantage of Theorem 1 is that the spectral projection matrices P_r, P_l have to be known, i.e. in general the matrices R, S. But then the decoupled system (15) is known also, and the asymptotic stability of system (4, 5) can be discussed by the slow subsystem $\dot{x}_1 = A_1 x_1$. This disadvantage can be reduced by using the quasi-Weierstrass form [1], such that a certain simplification is obtained. An additional simplification can be got by using orthogonal transformations resulting in a block triangle representation of the matrix pencil [4, 16, 17].

But for a simple comparison of the two approaches the transformation (13) is assumed in the following.

3.3 Second Approach

The second approach of the stability analysis of system (4, 5) is based on Eqs. (10, 11). They have been introduced to define 'admissible' systems which are asymptotically stable and 'impulse-free' ($N = 0$, $k = 1$) regular descriptor systems (4, 5), cf. [12, 15]. There is following result.

Theorem 2 *The regular descriptor system (4, 5) is admissible if and only if Eqs. (10, 11) can be solved for a positive definite matrix* $Q = Q^{\top} > 0$ *such that*

$$E^{\top} P_2 \geq 0 \tag{18}$$

holds.

Remark 3 The matrix Q in Theorem 2 has to be chosen suitably such that the left hand side of Eq. (11) can be considered in the sense of a linear matrix inequality [12, 15].

Remark 4 Theorem 2 provides a test of asymptotic stability of index-1-descriptor systems. In parallel it characterizes the index-1 property also. The disadvantage of Theorem 2 is that it does not cover higher index systems with $k \geq 2$. To overcome this disadvantage in the following the problem of asymptotic stability will be discussed on the basis of the Weierstrass canonical form (15) and the related Lyapunov equations (10, 11).

In [10] the Weierstrass canonical transformation (13) has been applied to Eqs. (10, 11). Then the canonical Lyapunov matrix equations of the second approach are obtained:
From (10) the symmetry condition reads as

$$\begin{bmatrix} I_1 & \\ & N^{\top} \end{bmatrix} \overline{P}_2 = \overline{P}_2^{\top} \begin{bmatrix} I_1 & \\ & N \end{bmatrix} \tag{19}$$

and from (11)

$$\begin{bmatrix} A_1^{\top} & \\ & I_2 \end{bmatrix} \overline{P}_2 + \overline{P}_2^{\top} \begin{bmatrix} A_1 & \\ & I_2 \end{bmatrix} = -\overline{Q} \tag{20}$$

is obtained where

$$\overline{P}_2 = S^{-\top} P_2 R, \quad \overline{Q} = R^{\top} Q R. \tag{21}$$

Introducing submatrices

$$\overline{P}_2 = \begin{bmatrix} P_{11} & P_{12} \\ P_{21} & P_{22} \end{bmatrix}, \quad \overline{Q} = \begin{bmatrix} Q_{11} & Q_{12} \\ Q_{12}^{\top} & Q_{22} \end{bmatrix} \tag{22}$$

Equation (19) results in

$$P_{11} = P_{11}^{\mathsf{T}}, \quad P_{12} = P_{21}^{\mathsf{T}} N, \quad N^{\mathsf{T}} P_{22} = P_{22}^{\mathsf{T}} N. \tag{23}$$

From the generalized Lyapunov equation (20) follows

$$A_1^{\mathsf{T}} P_{11} + P_{11} A_1 = -Q_{11}, \quad P_{21} + N^{\mathsf{T}} P_{21} A_1 = -Q_{12}^{\mathsf{T}}, \quad P_{22} + P_{22}^{\mathsf{T}} = -Q_{22}. \tag{24}$$

As mentioned above in Lemma 2 the stability behavior of (4, 5) is governed by the stability behavior of the slow subsystem $\dot{x}_1 = A_1 x_1$. Therefore, asymptotic stability of the descriptor system (4, 5) holds if and only if the corresponding first equations in (23) and (24) are satisfied with

$$P_{11} > 0, \quad Q_{11} > 0 \tag{25}$$

or in a weakened form, cf. [7]:

$$P_{11} > 0, \quad Q_{11} \geq 0, \quad (A_1, Q_{11}) \quad \text{completely observable.} \tag{26}$$

The other submatrices do not play a role and can be chosen arbitrarily satisfying (23, 24). The simplest choice is

$$Q_{12} = 0, \quad Q_{22} = 0, \quad P_{12} = 0, \quad P_{21} = 0, \quad P_{22} = 0. \tag{27}$$

By this choice asymptotic stability is given by

$$\overline{P}_2 = \overline{P}_2^{\mathsf{T}} \geq 0, \quad \operatorname{rk} \overline{P}_2 = n_1. \tag{28}$$

For the Weierstrass canonical representation the problem of asymptotic stability of (4, 5) is solved completely by (25–28). How this result can be transferred into the original representation (E, A, P, Q)? What conditions the matrix Q has to satisfy that $\overline{Q}$ has the form

$$\overline{Q} = R^{\mathsf{T}} Q R = \begin{bmatrix} Q_{11} & 0 \\ 0 & 0 \end{bmatrix}? \tag{29}$$

Similarly, what conditions P_2 has to satisfy that $\overline{P}_2$ has the form

$$\overline{P}_2 = S^{-\mathsf{T}} P_2 R = \begin{bmatrix} P_{11} & 0 \\ 0 & 0 \end{bmatrix}? \tag{30}$$

These questions have been discussed recently in [9] for a dual control problem to characterize proper or non-proper behavior of a controlled descriptor system. Applying this procedure to the problems (29, 30) the matrices

$$H = \begin{bmatrix} E^\top & & & \\ A^\top & E^\top & & \\ & \ddots & \ddots & \\ & & A^\top & E^\top \end{bmatrix}_{kn\times kn}, \quad T = \begin{bmatrix} & \vdots & Q \\ H & \vdots & 0 \\ & \vdots & \vdots \\ & \vdots & 0 \end{bmatrix} \tag{31}$$

have to be defined.

Lemma 3 *The matrix* $\overline{Q} = \overline{Q}^\top$ *has the form* (29) *if and only if*

$$\operatorname{rk} H = \operatorname{rk} T. \tag{32}$$

If $\operatorname{rk} T > \operatorname{rk} H$ then the second block row (and column) of $\overline{Q}$ is different from zero. Additionally, it is

$$\operatorname{rk} H = (k-1)n + n_1 = k\,n - n_2. \tag{33}$$

If the index k is not known in advance (by other considerations), then k can be determined by the sequence of ranks of the block principal submatrices of H [9]. These conditions have been proved in [9] for a dual control problem. Here, a different proof of (32) is shown.

Proof The condition (32) represents the requirement that

$$HX = \begin{bmatrix} Q \\ 0 \\ \vdots \\ 0 \end{bmatrix} \quad \text{with} \quad X = \begin{bmatrix} X_1 \\ X_2 \\ \vdots \\ X_k \end{bmatrix} \tag{34}$$

is solvable. Because regularity of A is a necessary condition for asymptotic stability of system (4, 5), Eq. (34) can be solved stepwise by

$$X_j = (-1)^{k-j}\left(A^{-\top}E^\top\right)^{k-j} X_k, \quad j = 1,\dots,k-1, \tag{35}$$

$$(-1)^{k-1}\, E^\top (A^{-\top}E^\top)^{k-1}\, X_k = Q. \tag{36}$$

To show the solvability of Eq. (36) (which is equivalent to the solvability of Eq. (34)), it will be transformed according the Weierstrass canonical form resulting in

$$(-1)^{k-1}\overline{E}^\top\left(\overline{A}^{-\top}\overline{E}^\top\right)\overline{X}_k = \overline{Q} \quad \text{with} \quad \overline{X}_k = S^{-\top}X_k\,R. \tag{37}$$

It is

$$\overline{\mathrm{E}}^{\top}\left(\overline{\mathrm{A}}^{-\top}\overline{\mathrm{E}}^{\top}\right)^{k-1} = \begin{bmatrix} \left(\mathrm{A}_1^{\top}\right)^{-(k-1)} & 0 \\ 0 & 0 \end{bmatrix}. \tag{38}$$

Therefore, Eq. (36) is solvable if and only if

$$n_1 = \mathrm{rk}\begin{bmatrix} \left(\mathrm{A}_1^{\top}\right)^{-(k-1)} & \mathrm{Q}_{11} & \mathrm{Q}_{12} \\ 0 & \mathrm{Q}_{12}^{\top} & \mathrm{Q}_{22} \end{bmatrix} \tag{39}$$

holds. This requirement is satisfied if and only if

$$\mathrm{Q}_{12} = 0, \quad \mathrm{Q}_{22} = 0 \tag{40}$$

showing the form (29) of $\overline{\mathrm{Q}}$.

Some additional information can be presented. Subdividing

$$\mathrm{S} = \begin{bmatrix} \mathrm{S}_1 \\ \mathrm{S}_2 \end{bmatrix}, \quad \mathrm{R} = \begin{bmatrix} \mathrm{R}_1 & \mathrm{R}_2 \end{bmatrix} \tag{41}$$

according to the two subsystems (15) then it follows from (30)

$$\mathrm{P}_2 = \mathrm{S}_1^{\top}\,\mathrm{P}_{11}\,\mathrm{S}_1\,\mathrm{E}, \quad \mathrm{R}_1^{\top}\left(\mathrm{E}^{\top}\,\mathrm{P}_2\right)\mathrm{R}_1 = \mathrm{P}_{11} > 0 \tag{42}$$

such that

$$\mathrm{E}^{\top}\mathrm{P}_2 = \mathrm{P}_2^{\top}\mathrm{E} = \mathrm{E}^{\top}\mathrm{S}_1\,\mathrm{P}_{11}\,\mathrm{S}_1\,\mathrm{E} \geq 0, \quad \mathrm{rk}\,\mathrm{E}^{\top}\mathrm{P}_2 = n_1. \tag{43}$$

Analogously

$$\mathrm{Q} = \mathrm{E}^{\top}\mathrm{S}_1^{\top}\,\mathrm{Q}_{11}\,\mathrm{S}_1\,\mathrm{E}, \quad \mathrm{R}_1^{\top}\mathrm{Q}\,\mathrm{R}_1 = \mathrm{Q}_{11} \tag{44}$$

is obtained. Summarizing the discussion the following is obtained.

Theorem 3 *The descriptor system (4, 5) is asymptotically stable if and only if there exist matrices* P_2, Q *satisfying Eqs. (10, 11)*

$$\mathrm{E}^{\top}\mathrm{P}_2 = \mathrm{P}_2^{\top}\mathrm{E} \qquad \textit{(symmetry)}$$

$$\mathrm{A}^{\top}\mathrm{P}_2 + \mathrm{P}_2^{\top}\mathrm{A} = -\mathrm{Q} \qquad \textit{(Lyaponov)}$$

with following properties:

$$\mathrm{Q} = \mathrm{Q}^{\top} \geq 0, \quad \mathrm{rk}\,\mathrm{Q} = n_1, \quad \mathrm{rk}\,\mathrm{H} = \mathrm{rk}\,\mathrm{T}, \tag{45}$$

$$\mathrm{E}^{\top}\mathrm{P}_2 \geq 0, \quad \mathrm{rk}\,\mathrm{P}_2 = \mathrm{rk}\,\mathrm{E}^{\top}\mathrm{P}_2 = n_1 \tag{46}$$

where $n_1 = \mathrm{rk}\,\mathrm{H} - (k-1)n$.

Remark 5 The condition $\mathrm{rk}\,\mathrm{Q} = n_1$ in (45) can be weakened into $\mathrm{rk}\,\mathrm{Q} \leq n_1$ with the so-called R-observability of (E, A; Q), cf. relations (26).

Remark 6 In principal, in the Theorems 1 and 3 the matrix Q has the same meaning. But in Theorem 1 every symmetric, positive definite matrix Q can be used, while in Theorem 3 the matrix Q has to satisfy the requirements (45). The main restriction is $\mathrm{rk}\,\mathrm{H} = \mathrm{rk}\,\mathrm{T}$ as it will be shown in Sect. 4.2 for holonomic mechanical systems.

The advantage of Theorem 3 is that the procedure operates with the original system data without applying any transformation (e.g. the Weierstrass canonical transformation). Nevertheless, the solution of (10, 11) and the checks of (45, 46) are expensive in general which will be shown by the example of a mechanical system with holonomic constraints.

4 Mechanical Descriptor Systems

According to Lagrange's equations of first kind the equations of motion of a linear time-invariant mechanical system with holonomic constraints are described by

$$\begin{aligned} \mathrm{M}\,\ddot{\mathrm{z}} + \mathrm{D}\,\dot{\mathrm{z}} + \mathrm{K}\,\mathrm{z} &= \mathrm{F}^{\mathsf{T}}\boldsymbol{\lambda}, \\ \mathrm{F}\,\mathrm{z} &= 0 \end{aligned} \tag{47}$$

where the various quantities mean

z ... vector of (redundant) displacements, $\dim \mathrm{z} = q$,

$\boldsymbol{\lambda}$ ... vector of Lagrange's multipliers, $\dim \boldsymbol{\lambda} = p$,

M ... matrix of inertia, $\dim \mathrm{M} = q \times q$, $\mathrm{M} = \mathrm{M}^{\mathsf{T}} > 0$,

D ... matrix of damping and gyroscopic forces, $\dim \mathrm{D} = q \times q$,

K ... matrix of stiffness and circulatory forces, $\dim \mathrm{K} = q \times q$,

F ... matrix of constraints, $\dim \mathrm{F} = p \times q$, $\mathrm{rk}\,\mathrm{F} = p$

The descriptor system notation according to (4) reads as

$$\mathrm{E}\,\dot{\mathrm{x}} = \mathrm{A}\,\mathrm{x}: \quad \mathrm{x} = \begin{bmatrix} \mathrm{z} \\ \dot{\mathrm{z}} \\ \boldsymbol{\lambda} \end{bmatrix},$$

$$\mathrm{E} = \begin{bmatrix} \mathrm{I}_q & & \\ & \mathrm{M} & \\ & & 0 \end{bmatrix}, \quad \mathrm{A} = \begin{bmatrix} 0 & \mathrm{I}_q & 0 \\ -\mathrm{K} & -\mathrm{D} & \mathrm{F}^{\mathsf{T}} \\ \mathrm{F} & 0 & 0 \end{bmatrix}. \tag{48}$$

The matrix pencil $(s\mathrm{E}-\mathrm{A})$ is regular, i.e. relation (5) is satisfied if

$$\mathrm{rk}\,\mathrm{M} = q, \quad \mathrm{rk}\,\mathrm{F} = p \tag{49}$$

hold which has been assumed.

Now, the stability analysis will be applied by both approaches of Sects. 3.2 and 3.3.

4.1 First Approach

In general, the transformation matrices R, S are not known explicitly, but for mechanical descriptor systems these matrices are available [13]. Defining an orthogonal complement L to F,

$$\dim \mathrm{L} = (q-p)\times q : \mathrm{rk}\begin{bmatrix}\mathrm{L}\\ \mathrm{F}\end{bmatrix} = q, \quad \mathrm{L}\,\mathrm{F}^{\mathsf{T}} = 0, \tag{50}$$

and introducing following matrices

$$\mathrm{U} = \mathrm{L}\,\mathrm{M}^{-1}\mathrm{W}, \quad \mathrm{V} = \left(\mathrm{F}\,\mathrm{M}^{-1}\mathrm{F}^{\mathsf{T}}\right)^{-1}\mathrm{F}\,\mathrm{M}^{-1}, \quad \mathrm{W} = \mathrm{I}_q - \mathrm{F}^{\mathsf{T}}\mathrm{V}, \tag{51}$$

then the transformation matrices R, S are given by

$$\mathrm{S} = \begin{bmatrix} \mathrm{L}\,\mathrm{W}^{\mathsf{T}} & 0 & \mathrm{U}\,\mathrm{D}\,\mathrm{V}^{\mathsf{T}} \\ \mathrm{U}\,\mathrm{D}\,\mathrm{V}^{\mathsf{T}}\mathrm{F} & \mathrm{U} & \mathrm{U}(\mathrm{K}-\mathrm{D}\,\mathrm{L}^{+}\mathrm{U}\,\mathrm{D})\mathrm{V}^{\mathsf{T}} \\ 0 & \mathrm{F}\,\mathrm{M}^{-1} & 0 \\ \mathrm{F} & 0 & 0 \\ 0 & 0 & \mathrm{I}_p \end{bmatrix} \tag{52}$$

$$\mathrm{R} = \begin{bmatrix} \mathrm{L}^{+} & 0 & 0 & 0 & \mathrm{V}^{\mathsf{T}} \\ 0 & \mathrm{L}^{+} & 0 & \mathrm{V}^{\mathsf{T}} & -\mathrm{L}^{+}\mathrm{U}\,\mathrm{D}\,\mathrm{V}^{\mathsf{T}} \\ \mathrm{V}\,\mathrm{K}\,\mathrm{L}^{+} & \mathrm{V}\,\mathrm{D}\,\mathrm{L}^{+} & (\mathrm{F}\,\mathrm{M}^{-1}\mathrm{F}^{\mathsf{T}})^{-1} & \mathrm{V}\,\mathrm{D}\,\mathrm{V}^{\mathsf{T}} & \mathrm{V}(\mathrm{K}-\mathrm{D}\,\mathrm{L}^{+}\mathrm{U}\,\mathrm{D})\mathrm{V}^{\mathsf{T}} \end{bmatrix} \tag{53}$$

$(\mathrm{L}^{+} = \mathrm{L}^{\mathsf{T}}(\mathrm{L}\,\mathrm{L}^{\mathsf{T}})^{-1})$. By that the spectral projection matrices (16) can be calculated explicitly:

$$\mathrm{P}_r = \begin{bmatrix} \mathrm{L}^{+}\mathrm{L} & 0 & 0 \\ \mathrm{L}^{+}\mathrm{U}\,\mathrm{D}\,\mathrm{V}^{\mathsf{T}}\mathrm{F} & \mathrm{L}^{+}\mathrm{L} & 0 \\ \mathrm{V}\mathrm{K}\mathrm{L}^{+}\mathrm{L} + \mathrm{V}\,\mathrm{D}\,\mathrm{L}^{+}\mathrm{U}\,\mathrm{D}\,\mathrm{V}^{\mathsf{T}}\mathrm{F} & \mathrm{V}\,\mathrm{D}\,\mathrm{L}^{+}\mathrm{L} & 0 \end{bmatrix}, \tag{54}$$

$$\mathrm{P}_l = \begin{bmatrix} \mathrm{L}^{+}\mathrm{L} & 0 & \mathrm{L}^{+}\mathrm{U}\,\mathrm{D}\,\mathrm{V}^{\mathsf{T}} \\ \mathrm{M}\,\mathrm{L}^{+}\mathrm{U}\,\mathrm{D}\,\mathrm{V}^{\mathsf{T}}\mathrm{F} & \mathrm{M}\,\mathrm{L}^{+}\mathrm{U} & \mathrm{M}\,\mathrm{L}^{+}\mathrm{U}(\mathrm{K}-\mathrm{D}\,\mathrm{L}^{+}\mathrm{U}\,\mathrm{D})\mathrm{V}^{\mathsf{T}} \\ 0 & 0 & 0 \end{bmatrix}. \tag{55}$$

Now, a suitable pair (P_1, Q) has to be found satisfying the generalized Lyapunov equation (17) of the first approach. After a laborious work the following result is obtained for a symmetric stiffness matrix $K = K^T$ (it is not surprising that for $K \neq K^T$ there was not a success because also for unconstrained mechanical systems this problem is very difficult to solve, cf. [7]):

$$Q = 2\begin{bmatrix} F^T V\, D^T U^T D_1 U\, D\, V^T F & F^T V\, D^T U^T D_1 L\, W^T & 0 \\ W\, L^T D_1 U\, D\, V^T F & W\, L^T D_1 L\, W^T & 0 \\ 0 & 0 & 0 \end{bmatrix}, \tag{56}$$

$P_1 = [\overline{P}_{ij}]$ with

$$\begin{aligned} \overline{P}_{11} &= W\, L^T K_1 L\, W^T + F^T V\, D^T U^T M_1 U\, D\, V^T F \\ \overline{P}_{12} &= F^T V\, D^T U^T M_1 U \\ \overline{P}_{13} &= W\, L^T K_1 U\, D\, V^T + F^T V\, D^T U^T M_1 U(K - D\, L^+ U\, D)V^T \\ \overline{P}_{22} &= U^T M_1 U \\ \overline{P}_{23} &= U^T M_1 U(K - D\, L^+ U\, D)V^T \\ \overline{P}_{33} &= V\, D U^T K_1 U\, D\, V^T + V(K - D^T U^T L^{+T} D^T)U^T M_1 U(K - D\, L^+ U\, D)V^T \end{aligned} \tag{57}$$

where

$$\begin{aligned} M_1 &= L^{+T} M\, L^+, & K_1 &= K_1^T = L^{+T} K L^+, \\ D_1 + G_1 &= L^{+T} D\, L^+, & D_1 &= D_1^T, \quad G_1 = -G_1^T. \end{aligned} \tag{58}$$

Additionally to Eq. (17) the required definiteness properties have to be satisfied. Q is a positive semidefinite matrix if D_1 is positive semidefinite:

$$Q = [\overline{Q}_{ij}] : \begin{bmatrix} \overline{Q}_{11} & \overline{Q}_{12} \\ \overline{Q}_{12}^T & \overline{Q}_{22} \end{bmatrix} = Z^T \begin{bmatrix} 0 & 0 \\ 0 & 2D_1 \end{bmatrix} Z \tag{59}$$

where

$$Z = \begin{bmatrix} L\, W^T & 0 \\ U\, D\, V^T F & L\, W^T \end{bmatrix}. \tag{60}$$

Also P_1 is a positive semidefinite matrix if K_1 is positive definite:

$$P_1 = S_1^T \overline{\overline{P}}_{11} S_1 \geq 0 \text{ with } \overline{\overline{P}}_{11} = \begin{bmatrix} K_1 & 0 \\ 0 & M_1 \end{bmatrix} > 0, \tag{61}$$

$$S_1 = \begin{bmatrix} L\,W^T & 0 & U\,D\,V^T \\ U\,D\,V^T F & U & U(K - D\,L^+ U\,D)V^T \end{bmatrix} \tag{62}$$

The assumption of Theorem 1 on $Q = Q^T > 0$ is not met because it is $Q = Q^T \geq 0$ only. But according to Remark 1 the Theorem 1 holds for semidefinite Q if additionally the R-observability of $(E, A; Q)$ is satisfied. Therefore, to guarantee asymptotic stability the conditions

$$M_1 > 0, \quad D_1 \geq 0, \quad K_1 > 0 \tag{63}$$

are sufficient together with the R-observability. In mechanical sciences this observability condition is called the condition of 'pervasive damping', cf. [7].

The Lyapunov function (6) represents the total energy of the constrained mechanical system. Its time-derivate is twice the dissipation function. By this mechanical background it was possible to construct the matrices P_1 and Q of the first approach.

Finally, Lemma 1 is applied to prepare a result for the second approach in the next section:

$$P_2 = P_1 E = \begin{bmatrix} \overline{P}_{11} & \overline{P}_{12} M & 0 \\ \overline{P}_{12}^T & \overline{P}_{22} M & 0 \\ \overline{P}_{13}^T & \overline{P}_{23}^T M & 0 \end{bmatrix} \tag{64}$$

If for the second approach the matrix Q (56) can be used for the stability analysis, then P_2 (64) represents a solution of (10, 11).

4.2 Second Approach

The general result of Theorem 3 is applied to the holonomic mechanical system (47). The index is $k = 3$. Therefore, the rank condition (32), $\text{rk}\,H = \text{rk}\,T$, has to be checked by superblockmatrices with 3 block rows. Assuming

$$P_2 = [P_{ij}], \quad Q = Q^T = [Q_{ij}], \qquad i, j = 1, 2, 3, \tag{65}$$

then after some costly manipulations of the matrices H, T the relation

$$\mathrm{rk\,T} = \mathrm{rk\,H} + \mathrm{rk}\begin{bmatrix} Q_{13}^T & Q_{23}^T & Q_{33} \\ -F M^{-1} Q_{12}^T & -F M^{-1} Q_{22} & -F M^{-1} Q_{23} \\ F M^{-1}(Q_{11} - D^T M^{-1} Q_{12}^T) & F M^{-1}(Q_{12} - D^T M^{-1} Q_{22}) & F M^{-1}(Q_{13} - D^T M^{-1} Q_{23}) \end{bmatrix} \tag{66}$$

is obtained. This results for $\mathrm{rk\,H} = \mathrm{rk\,T}$ in

$$Q_{13} = 0, \quad Q_{23} = 0, \quad Q_{33} = 0, \tag{67}$$

$$F M^{-1} Q_{12}^T = 0, \quad F M^{-1} Q_{22} = 0, \tag{68}$$

$$F M^{-1}(Q_{11} - D^T M^{-1} Q_{12}^T) = 0, \quad F M^{-1}(Q_{12} - D^T M^{-1} Q_{22}) = 0, \tag{69}$$

Therefore, Q has the desired form

$$Q = Q^T = \begin{bmatrix} Q_{11} & Q_{12} & 0 \\ Q_{12}^T & Q_{22} & 0 \\ 0 & 0 & 0 \end{bmatrix} \tag{70}$$

where Q_{11}, Q_{12}, Q_{22} still have to satisfy Eqs. (68, 69).
The symmetry condition (10) leads to the requirements

$$P_{11} = P_{11}^T, \quad P_{12} = P_{21}^T\, M, \quad P_{13} = 0, \quad P_{22}^T\, M = M\, P_{22} \quad P_{23} = 0 \tag{71}$$

The generalized Lyapunov matrix equation (11) results in

$$-K^T P_{21} - P_{21}^T\, K + F^T P_{31} + P_{31}^T\, F = -Q_{11}, \tag{72}$$

$$-K^T P_{22} + F^T P_{32} + P_{11} - P_{21}^T\, D = -Q_{12}, \tag{73}$$

$$F^T P_{33} + P_{21}^T\, F^T = 0, \tag{74}$$

$$P_{21}^T\, M + M\, P_{21} - D^T P_{22} - P_{22}\, D = -Q_{22}, \tag{75}$$

$$F\, P_{22} = 0. \tag{76}$$

Again, only for a symmetric stiffness matrix $K = K^T$ it is possible to solve Eqs. (71–76) for the matrix Q (56) introduced above in Sect. 4.1. Then the solution for P_2 is exactly how it was expected by (64):

$$P_2 = \begin{bmatrix} \overline{P}_{11} & \overline{P}_{12}\, M & 0 \\ \overline{P}_{12}^T & \overline{P}_{22}\, M & 0 \\ \overline{P}_{13}^T & \overline{P}_{23}^T\, M & 0 \end{bmatrix} = P_1\, E, \tag{77}$$

For the stability analysis the matrices Q, $E^{\mathsf{T}} P_2$ should be positive semidefinite at least. The check of $Q = Q^{\mathsf{T}} \geq 0$ has been done by (59) and is guaranteed for $D_1 \geq 0$. The check of $E^{\mathsf{T}} P_2$ leads to

$$E^{\mathsf{T}} P_2 = \begin{bmatrix} \overline{P}_{11} & \overline{P}_{12}\, M & 0 \\ M\, \overline{P}_{12}^{\mathsf{T}} & M\, \overline{P}_{22}\, M & 0 \\ 0 & 0 & 0 \end{bmatrix} = \begin{bmatrix} Z^{\mathsf{T}} \begin{bmatrix} K_1 & 0 \\ 0 & M_1 \end{bmatrix} Z & \vdots & 0 \\ & \vdots & 0 \\ \cdots \quad \cdots \quad \cdots & \vdots & \cdots \\ 0 \ \cdots \ 0 & \vdots & 0 \end{bmatrix} \tag{78}$$

such that $\operatorname{rk} E^{\mathsf{T}} P_2 = n_1 = 2(q-p)$ and $E^{\mathsf{T}} P_2 \geq 0$ for $K_1 > 0$.

The statement of the second approach is exactly the same as of the first approach. Only the representation of the solutions is different. All over, the mechanical system with holonomic constraints (47) is asymptotically stable for a positive definite stiffness matrix $K_1 = L^{+\mathsf{T}} K\, L^{+} > 0$ ($K = K^{\mathsf{T}}$) and pervasive damping $D_1 \geq 0$, i.e. (E, A; Q)-R-observability.

5 Comparisation of the Two Approaches

At a first view the two approaches look differently. The examples in Sect. 2 show that the generalized Lyapunov equations (7, 8) and (10, 11) may have completely different solution behavior. But confining oneself to the problem of asymptotic stability of the descriptor system (4, 5), specializing (7, 8) to (17), and introducing the requirement (32), then both approaches concentrate totally on the slow subsystem $\dot{x}_1 = A_1 x_1$. The fast subsystem $N\dot{x}_2 = x_2$ is suppressed completely. If the slow subsystem is asymptotically stable, cf. Lemma 2, then both approaches result in uniquely determined solutions P_1 and P_2 which are related by Eq. (12). In this sense the two approaches coincide for asymptotic stability. The main difference between the two approaches consists in the type of processing. The first approach needs data of the Weierstrass canonical form which usually are hard to obtain. The second approach uses only data of the original system and can be applied generally.

Conclusions
The stability analysis of linear time-invariant descriptor system has been considered by two different types of generalized Lyapunov equations. Additionally to the first approach represented in [14] a new, second approach has been shown. In case of asymptotic stability both approaches are equivalent but the second approach has the advantage that all calculations can be realized

(continued)

by the original system data without any transformation into the Weierstrass canonical form. Nevertheless, the example of holonomic mechanical systems has shown that both approaches are very costly in general.

References

1. Berger, T., Ilchmann, A., Trenn, S.: The quasi-Weierstraß form for regular matrix pencils. Linear Algebra Appl. **436**, 4052–069 (2012)
2. Dai, L.: Singular Control Systems. Lecture Notes in Control and Information Sciences, vol. 118. Springer, Berlin (1989)
3. Duan, G.-R.: Analysis and Design of Descriptor Linear Systems. Advances in Mechanics and Mathematics, vol. 23, Springer, New York (2010)
4. Hou, M.: Descriptor Systems: Observers and Fault Diagnosis. Fortschr.-Ber. VDI, Reihe 8, Nr. 482. VDI, Düsseldorf (1995)
5. Ishihara, J.Y., Terra, M.H.: On the Lyapunov theorem for singular systems. IEEE Trans. Autom. Control **47**, 1926–1930 (2002)
6. Mehrmann, V., Stykel, T.: Balanced truncation model reduction for large-scale systems in descriptor form. In: Benner, P., Mehrmann, V., Sorenson, D.C. (eds.) Dimension Reduction of Large-Scale Systems. Lecture Notes in Computational Science and Engineering, vol. 45, pp. 83–115. Springer, Berlin (2005)
7. Müller, P.C.: Stabilität und Matrizen. Springer, Berlin (1977)
8. Müller, P.C.: Stability of linear mechanical systems with holonomic constraints. Appl. Mech. Rev. **46**(11, part 2), 160–164 (1993)
9. Müller, P.C.: Characteristics of LTI descriptor systems. PAMM-Proc. Appl. Math. Mech. **11**, 831–832 (2011)
10. Müller, P.C.: Generalized Lyapunov matrix equation revisited. In: Proceedings of the XV International Symppsium on Dynamic Problems of Mechanics (DINAME 2013), Buzios RJ, Brazil (2013)
11. Owens, D.H., Debeljkovic, D.L.: Consistency and Liapunov stability of linear descriptor systems: a geometric analysis. IMA J. Math. Control Inf. **2**, 139–151 (1985)
12. Rehm, A.: Control of Linear Descriptor Systems: A Matrix Inequality Approach. Fortschr.-Ber. VDI, Reihe 8, Nr. 1019. VDI, Düsseldorf (2004)
13. Schüpphaus, R.: Regelungstechnische Analyse und Synthese von Mehrkörpersystemen in Deskriptorform. Fortschr.-Ber. VDI, Reihe 8, Nr. 478. VDI, Düsseldorf (1995)
14. Stykel, T.: Analysis and numerical solution of generalized Lyapunov equations. Ph.D. thesis, Institute for Mathematics, TU Berlin (2002)
15. Tabaka, K., Morihira, N., Katayama, T.: A generalized Lyapunov theorem for descriptor systems. Syst. Control Lett. **24**, 49–51 (1995)
16. Van Dooren, P.: The computation of Kronecker's canonical form of a singular pencil. Linear Algebra Appl. **27**, 103–140 (1979)
17. Van Dooren, P., Dewilde, P.: The eigenstructure of an arbitrary polynomial matrix: Computational aspects. Linear Algebra Appl. **50**, 545–579 (1983)

Zero Dynamics and Stabilization for Linear DAEs

Thomas Berger

Abstract We study linear differential-algebraic multi-input multi-output systems which are not necessarily regular and investigate the asymptotic stability of the zero dynamics and stabilizability. To this end, the concepts of autonomous zero dynamics, transmission zeros, right-invertibility, stabilizability in the behavioral sense and detectability in the behavioral sense are introduced and algebraic characterizations are derived. It is then proved, for the class of right-invertible systems with autonomous zero dynamics, that asymptotic stability of the zero dynamics is equivalent to three conditions: stabilizability in the behavioral sense, detectability in the behavioral sense, and the condition that all transmission zeros of the system are in the open left complex half-plane. Furthermore, for the same class, it is shown that we can achieve, by a compatible control in the behavioral sense, that the Lyapunov exponent of the interconnected system equals the Lyapunov exponent of the zero dynamics.

Keywords Differential-algebraic equations • Zero dynamics • Transmission zeros • Right-invertibility • Stabilizability • Detectability • Lyapunov exponent

Mathematics Subject Classification (2010) 34A09 · 15A22 · 93B05 · 93B07 · 93B25

The author was supported by DFG grant IL 25/9.

T. Berger (✉)
Fachbereich Mathematik, Universität Hamburg, Bundesstraße 55, 20146 Hamburg, Germany
e-mail: thomas.berger@uni-hamburg.de

S. Schöps et al. (eds.), *Progress in Differential-Algebraic Equations*,
Differential-Algebraic Equations Forum, DOI 10.1007/978-3-662-44926-4_2

Nomenclature

$\mathbb{N}$, $\mathbb{N}_0$		the set of natural numbers, $\mathbb{N}_0 = \mathbb{N} \cup \{0\}$
$L(\alpha), \lvert\alpha\rvert$		length $L(\alpha) = l$ and absolute value $\lvert\alpha\rvert = \sum_{i=1}^{l} \alpha_i$ of a multi-index $\alpha = (\alpha_1, \ldots, \alpha_l) \in \mathbb{N}^l$
$\mathbb{C}_+(\mathbb{C}_-)$		open set of complex numbers with positive (negative) real part, resp.
$\mathbb{R}[s]$		the ring of polynomials with coefficients in $\mathbb{R}$
$\mathbb{R}(s)$		the quotient field of $\mathbb{R}[s]$
$R^{n\times m}$		the set of $n \times m$ matrices with entries in a ring R
$\mathbf{Gl}_n(R)$		the group of invertible matrices in $R^{n\times n}$
$\sigma(M)$		the spectrum of $M \in \mathbb{R}^{n\times m}$
$\lVert x\rVert$	$=$	$\sqrt{x^\top x}$, the Euclidean norm of $x \in \mathbb{R}^n$
$\lVert M\rVert$	$=$	$\max\left\{\lVert M x\rVert \,\middle|\, x \in \mathbb{R}^m,\ \lVert x\rVert = 1\right\}$, induced norm of $M \in \mathbb{R}^{n\times m}$
$M^{-1}\mathcal{S}$	$=$	$\{\, x \in \mathbb{R}^m \mid Mx \in \mathcal{S} \,\}$, the pre-image of the set $\mathcal{S} \subseteq \mathbb{R}^n$ under $M \in \mathbb{R}^{n\times m}$
$\mathcal{C}^\infty(\mathbb{R};\mathbb{R}^n)$		the set of infinitely-times continuously differentiable functions $f : \mathbb{R} \to \mathbb{R}^n$
$\mathcal{L}^1_{\mathrm{loc}}(\mathbb{R};\mathbb{R}^n)$		the set of locally Lebesgue integrable functions $f : \mathbb{R} \to \mathbb{R}^n$, where $\int_K \lVert f(t)\rVert\, \mathrm{d}t < \infty$ for all compact $K \subseteq \mathbb{R}$
$\dot f\ (f^{(i)})$		the (i-th) weak derivative of $f \in \mathcal{L}^1_{\mathrm{loc}}(\mathbb{R};\mathbb{R}^n)$, $i \in \mathbb{N}_0$, see [1, Chap. 1]
$\mathcal{W}^{k,1}_{\mathrm{loc}}(\mathbb{R};\mathbb{R}^n)$	$=$	$\left\{\, f \in \mathcal{L}^1_{\mathrm{loc}}(\mathbb{R};\mathbb{R}^n) \,\middle|\, f^{(i)} \in \mathcal{L}^1_{\mathrm{loc}}(\mathbb{R};\mathbb{R}^n),\ i = 0, \ldots, k \,\right\}, k \in \mathbb{N}_0$
$f \overset{\text{a.e.}}{=} g$		means that $f, g \in \mathcal{L}^1_{\mathrm{loc}}(\mathbb{R};\mathbb{R}^n)$ are equal "almost everywhere", i.e., $f(t) = g(t)$ for almost all (a.a.) $t \in \mathbb{R}$
$\text{ess-sup}_I \lVert f\rVert$		the essential supremum of the measurable function $f : \mathbb{R} \to \mathbb{R}^n$ over $I \subseteq \mathbb{R}$
$f\vert_I$		the restriction of the function $f : \mathbb{R} \to \mathbb{R}^n$ to $I \subseteq \mathbb{R}$

1 Introduction

We consider linear constant coefficient DAEs of the form

$$\begin{aligned} \tfrac{\mathrm{d}}{\mathrm{d}t}Ex(t) &= Ax(t) + Bu(t) \\ y(t) &= Cx(t), \end{aligned} \tag{1}$$

where $E, A \in \mathbb{R}^{\ell \times n}$, $B \in \mathbb{R}^{\ell \times m}$, $C \in \mathbb{R}^{p \times n}$. The set of these systems is denoted by $\Sigma_{\ell,n,m,p}$ and we write $[E, A, B, C] \in \Sigma_{\ell,n,m,p}$. In the present paper, we put special emphasis on the non-regular case, i.e., we do not assume that $sE - A$ is *regular*, which would mean that $\ell = n$ and $\det(sE - A) \in \mathbb{R}[s] \setminus \{0\}$.

The functions $u : \mathbb{R} \to \mathbb{R}^m$ and $y : \mathbb{R} \to \mathbb{R}^p$ are called *input* and *output* of the system, resp. A trajectory $(x, u, y) : \mathbb{R} \to \mathbb{R}^n \times \mathbb{R}^m \times \mathbb{R}^p$ is said to be a *solution* of (1) if, and only if, it belongs to the *behavior* of (1):

$$\mathfrak{B}_{(1)} := \left\{ (x, u, y) \in \mathscr{L}^1_{\mathrm{loc}}(\mathbb{R}; \mathbb{R}^n \times \mathbb{R}^m \times \mathbb{R}^p) \;\middle|\; \begin{array}{l} Ex \in \mathscr{W}^{1,1}_{\mathrm{loc}}(\mathbb{R}; \mathbb{R}^\ell) \text{ and } (x, u, y) \\ \text{solves (1) for a.a. } t \in \mathbb{R} \end{array} \right\}.$$

Recall that any function $z \in \mathscr{W}^{1,1}_{\mathrm{loc}}(\mathbb{R}; \mathbb{R}^\ell)$ is in particular continuous.

Particular emphasis is placed on the *zero dynamics* of (1). These are, for $[E, A, B, C] \in \Sigma_{\ell,n,m,p}$, defined by

$$\mathscr{Z}\mathscr{D}_{(1)} := \left\{ (x, u, y) \in \mathfrak{B}_{(1)} \;\middle|\; y \overset{\text{a.e.}}{=} 0 \right\}.$$

By linearity of (1), $\mathscr{Z}\mathscr{D}_{(1)}$ is a real vector space.

The zero dynamics of (1) are called *autonomous* if, and only if,

$$\forall\, w_1, w_2 \in \mathscr{Z}\mathscr{D}_{(1)}\ \forall\, I \subseteq \mathbb{R} \text{ open interval}: \quad w_1|_I \overset{\text{a.e.}}{=} w_2|_I \implies w_1 \overset{\text{a.e.}}{=} w_2;$$

and *asymptotically stable* if, and only if,

$$\forall\, (x, u, y) \in \mathscr{Z}\mathscr{D}_{(1)}: \quad \lim_{t \to \infty} \operatorname{ess-sup}_{[t,\infty)} \|(x, u)\| = 0.$$

Note that the above definitions are within the spirit of the *behavioral approach* [20] and take into account that the zero dynamics $\mathscr{Z}\mathscr{D}_{(1)}$ are a linear behavior. In this framework the definition for autonomy of a general behavior is given in [20, Sec. 3.2] and the definition of asymptotic stability in [20, Def. 7.2.1].

(Asymptotically stable) zero dynamics are the vector space of those trajectories of the system which are, loosely speaking, not visible at the output (and tend to zero).

In the present paper, we show for the class of right-invertible systems with autonomous zero dynamics, that asymptotic stability of the zero dynamics is equivalent to the three conditions: stabilizability in the behavioral sense, detectability in the behavioral sense and the condition that all transmission zeros are in the open left complex half-plane. Furthermore, we show that we can achieve, by a compatible control in the behavioral sense, that the Lyapunov exponent of the interconnected system equals the Lyapunov exponent of the zero dynamics. In Sect. 2 we collect some basic control theoretic concepts such as transmission zeros, right-invertibility, stabilizability in the behavioral sense and detectability in the behavioral sense, and give algebraic characterizations of them. The first main result of the present paper, namely Theorem 1, is then stated and proved in Sect. 3 and some consequences for regular systems are derived. In Sect. 4 we introduce the concepts of compatible control (in the behavioral sense) and Lyapunov exponent for DAE systems and prove the second main result, namely Theorem 2.

For the application of compatible control it is necessary that the states and inputs of the DAE system are fixed a priori by the designer in order to establish the control law. This is different from other approaches based on the behavioral setting, see [12], where only the free variables in the system are viewed as inputs; this may require a reinterpretation of states as inputs and of inputs as states. In the present paper we will assume that such a reinterpretation of variables has already been done or is not feasible, and the given DAE system is fix.

2 Some Control Theoretic Concepts

In this section we recall the concepts used in the present paper in a control theoretic way and give useful algebraic characterizations. These concepts include transmission zeros, right-invertibility, stabilizability in the behavioral sense and detectability in the behavioral sense. We start with characterizations of autonomous and asymptotically stable zero dynamics, which have been introduced in Sect. 1.

Lemma 1 (Autonomous and stable zero dynamics) *Let $[E, A, B, C] \in \Sigma_{\ell,n,m,p}$. Then we have the following equivalences:*

(i) $\mathcal{Z}\mathcal{D}_{(1)}$ *are autonomous* $\iff \operatorname{rk}_{\mathbb{R}[s]} \begin{bmatrix} sE - A & -B \\ -C & 0 \end{bmatrix} = n + m.$

(ii) $\mathcal{Z}\mathcal{D}_{(1)}$ *are asymptotically stable* $\iff \forall\, \lambda \in \overline{\mathbb{C}}_+ : \operatorname{rk}_{\mathbb{C}} \begin{bmatrix} \lambda E - A & -B \\ -C & 0 \end{bmatrix} = n + m.$

Proof (i) follows from [4, Prop. 4.1.5] and (ii) from [4, Lem. 4.3.9]. □

Note that the above cited results from [4] have been first reported in [5]; in the following, this holds true for all results cited from [4].

The autonomy of the zero dynamics allows for a decomposition of the system, provided that C has full row rank. The main result of the present paper (see Sect. 3) is based on this decomposition.

Lemma 2 (System decomposition) *Let $[E, A, B, C] \in \Sigma_{\ell,n,m,p}$ with autonomous zero dynamics and* $\operatorname{rk} C = p$. *Then there exist $S \in \mathbf{Gl}_\ell(\mathbb{R})$ and $T \in \mathbf{Gl}_n(\mathbb{R})$ such that*

$$\begin{bmatrix} s\hat{E} - \hat{A} & \hat{B} \\ \hat{C} & 0 \end{bmatrix} = \begin{bmatrix} S & 0 \\ 0 & I_p \end{bmatrix} \begin{bmatrix} sE - A & B \\ C & 0 \end{bmatrix} \begin{bmatrix} T & 0 \\ 0 & I_m \end{bmatrix}, \tag{2}$$

where

$$\hat{E} = \begin{bmatrix} I_k & 0 & 0 \\ 0 & E_{22} & E_{23} \\ 0 & E_{32} & N \\ 0 & E_{42} & E_{43} \end{bmatrix}, \quad \hat{A} = \begin{bmatrix} Q & A_{12} & 0 \\ A_{21} & A_{22} & 0 \\ 0 & 0 & I_{n_3} \\ 0 & A_{42} & 0 \end{bmatrix}, \quad \hat{B} = \begin{bmatrix} 0 \\ I_m \\ 0 \\ 0 \end{bmatrix}, \quad \hat{C} = [0, I_p, 0], \tag{3}$$

$$k = \dim \mathcal{Z}\mathcal{D}_{(1)}, \tag{4}$$

and $N \in \mathbb{R}^{n_3 \times n_3}$, $n_3 = n - k - p$, is nilpotent with $N^\nu = 0$ and $N^{\nu-1} \neq 0$, $\nu \in \mathbb{N}$, $E_{22}, A_{22} \in \mathbb{R}^{m \times p}$ and all other matrices are of appropriate sizes.

Proof This result can be found in [4, Thm. 4.2.7]. □

An important characterization of asymptotically stable zero dynamics is the following, which is from [4, Cor. 4.2.11].

Lemma 3 (Stable zero dynamics) *Let $[E, A, B, C] \in \Sigma_{\ell,n,m,p}$ with autonomous zero dynamics and* $\operatorname{rk} C = p$. *Then, using the notation from Lemma 2, the zero dynamics $\mathcal{Z}\mathcal{D}_{(1)}$ are asymptotically stable if, and only if, $\sigma(Q) \subseteq \mathbb{C}_-$.*

Next, in order to define transmission zeros, we introduce the Smith-McMillan form of a rational matrix function.

Definition 1 (Smith-McMillan form [16, Sec. 6.5.2]) Let $G(s) \in \mathbb{R}(s)^{m \times p}$ with $\operatorname{rk}_{\mathbb{R}(s)} G(s) = r$. Then there exist $U(s) \in \mathbf{Gl}_m(\mathbb{R}[s])$, $V(s) \in \mathbf{Gl}_p(\mathbb{R}[s])$ such that

$$U(s)G(s)V(s) = \operatorname{diag}\left(\frac{\varepsilon_1(s)}{\psi_1(s)}, \ldots, \frac{\varepsilon_r(s)}{\psi_r(s)}, 0_{(m-r)\times(p-r)} \right),$$

where $\varepsilon_i(s), \psi_i(s) \in \mathbb{R}[s]$ are monic, coprime and satisfy $\varepsilon_i(s) | \varepsilon_{i+1}(s)$, $\psi_{i+1}(s) | \psi_i(s)$ for $i = 1, \ldots, r-1$. The number $s_0 \in \mathbb{C}$ is called *zero* of $G(s)$ if, and only if, $\varepsilon_r(s_0) = 0$ and *pole* of $G(s)$ if, and only if, $\psi_1(s_0) = 0$.

In the following we give the definition of transmission zeros for the system $[E, A, B, C]$. In fact, there are many different possibilities to define transmission zeros of control systems, even in the ODE case, see [13]; and they are not equivalent. We follow the definition given by Rosenbrock [21]: For $[I, A, B, C] \in \Sigma_{n,n,m,p}$, the transmission zeros are the zeros of the transfer function $C(sI - A)^{-1}B$. This definition has been generalized to regular DAE systems with transfer function $C(sE - A)^{-1}B$ in [7, Def. 5.3]. In the present framework, we do not require regularity of $sE - A$ and so a transfer function does in general not exist. However, it is possible to give a generalization of the inverse transfer function if the zero dynamics of $[E, A, B, C] \in \Sigma_{\ell,n,m,p}$ are autonomous: Let $L(s)$ be a left inverse of $\left[\begin{smallmatrix} sE-A & -B \\ -C & 0 \end{smallmatrix}\right]$ over $\mathbb{R}(s)$ (which exists by Lemma 1) and define

$$H(s) := -[0, I_m]L(s)\begin{bmatrix} 0 \\ I_p \end{bmatrix} \in \mathbb{R}(s)^{m\times p}. \tag{5}$$

It can be shown that $H(s)$ is independent of the choice of the left inverse $L(s)$ [4, Lem. 4.3.2] and if $sE-A$ is regular and $m = p$, then $H(s) = \big(C(sE-A)^{-1}B\big)^{-1}$ [4, Rem. 4.3.3], i.e., $H(s)$ is indeed the inverse of the transfer function in case of regularity. The fact that the zeros of $H(s)^{-1}$ are the poles of $H(s)$ and vice versa motivates the following definition.

Definition 2 (Transmission zeros) Let $[E, A, B, C] \in \Sigma_{\ell,n,m,p}$ with autonomous zero dynamics. Let $L(s)$ be a left inverse of $\left[\begin{smallmatrix} sE-A & -B \\ -C & 0 \end{smallmatrix}\right]$ over $\mathbb{R}(s)$ and let $H(s)$ be given as in (5). Then $s_0 \in \mathbb{C}$ is called *transmission zero* of $[E, A, B, C]$ if, and only if, s_0 is a pole $H(s)$.

Now we recall the definition of right-invertibility of a system from [22, Sec. 8.2].

Definition 3 (Right-invertibility) $[E, A, B, C] \in \Sigma_{\ell,n,m,p}$ is called *right-invertible* if, and only if,

$$\forall\, y \in \mathscr{C}^\infty(\mathbb{R};\mathbb{R}^p)\ \exists\, (x,u) \in \mathscr{L}^1_{\text{loc}}(\mathbb{R};\mathbb{R}^n) \times \mathscr{L}^1_{\text{loc}}(\mathbb{R};\mathbb{R}^m) :\ (x,u,y) \in \mathfrak{B}_{(1)}.$$

Right-invertibility may be characterized for systems with autonomous zero dynamics in terms of the form (3).

Lemma 4 (Right-invertibility and system decomposition) *Let $[E, A, B, C] \in \Sigma_{\ell,n,m,p}$ with autonomous zero dynamics. Then, using the notation from Lemma 2,*

$$[E, A, B, C] \text{ is right-invertible} \quad \Longleftrightarrow \quad \begin{cases} \operatorname{rk} C = p,\ E_{42} = 0,\ A_{42} = 0 \text{ and} \\ E_{43}N^j E_{32} = 0 \text{ for } j = 0, \ldots, \nu - 1. \end{cases}$$

Proof A proof can be found in [4, Prop. 4.2.12]. □

We are now in a position to characterize the transmission zeros in terms of the form (3).

Corollary 1 (Transmission zeros in decomposition) *Let $[E, A, B, C] \in \Sigma_{\ell,n,m,p}$ be right-invertible and have autonomous zero dynamics. Let $L(s)$ be a left inverse of $\left[\begin{smallmatrix} sE-A & -B \\ -C & 0 \end{smallmatrix}\right]$ over $\mathbb{R}(s)$ and let $H(s)$ be given as in (5). Then, using the notation from Lemma 2,*

$$H(s) = sE_{22} - A_{22} - A_{21}(sI_k - Q)^{-1}A_{12} - s^2E_{23}(sN - I_{n_3})^{-1}E_{32}$$

and $s_0 \in \mathbb{C}$ is a transmission zero of $[E, A, B, C]$ if, and only if, s_0 is a pole of

$$A_{21}(sI_k - Q)^{-1}A_{12}.$$

Proof The representation of $H(s)$ follows from [4, Lem. 4.3.2] and the characterization of transmission zeros is then immediate since $sE_{22} - A_{22} - s^2E_{23}(sN - I)^{-1}E_{32}$ is a polynomial as N is nilpotent and hence

$$(sN - I)^{-1} = -I - sN - \ldots - s^{\nu-1}N^{\nu-1}. \tag{6}$$

□

In the remainder of this section we introduce and characterize the concepts of stabilizability and detectability in the behavioral sense. (Behavioral) stabilizability for systems $[E, A, B, C] \in \Sigma_{\ell,n,m,p}$ is well-investigated, see e.g. the survey [8]. Detectability has been first defined and characterized for regular systems in [2]. For general DAE systems, a definition and characterization can be found in [14]; see also the equivalent definition in [20, Sec. 5.3.2]. The latter definition is given within the behavioral framework, however it is yet too restrictive for our purposes and it is not dual to the respective stabilizability concept. We use the following concepts of behavioral stabilizability and detectability.

Definition 4 (Stabilizability and detectability) $[E, A, B, C] \in \Sigma_{\ell,n,m,p}$ is called

(i) *stabilizable in the behavioral sense* if, and only if,

$$\forall\, (x, u, y) \in \mathfrak{B}_{(1)}\ \exists\, (x_0, u_0, y_0) \in \mathfrak{B}_{(1)} :$$
$$\big(\forall\, t < 0 : \ (x(t), u(t)) = (x_0(t), u_0(t))\big) \wedge \lim_{t\to\infty} \text{ess-sup}_{[t,\infty)} \|(x_0, u_0)\| = 0.$$

(ii) *detectable in the behavioral sense* if, and only if,

$$\forall\, (x, 0, 0) \in \mathfrak{B}_{(1)}\ \exists\, (x_0, 0, 0) \in \mathfrak{B}_{(1)} :$$
$$(\forall\, t < 0 : \ x(t) = x_0(t)) \ \wedge \ \lim_{t\to\infty} \text{ess-sup}_{[t,\infty)} \|x_0\| = 0.$$

In order to derive duality of the above concepts it is useful to consider, for $E, A \in \mathbb{R}^{\ell\times n}$, the DAE

$$\tfrac{\mathrm{d}}{\mathrm{d}t}Ex(t) = Ax(t) \tag{7}$$

without inputs and outputs. The behavior of (7) is given by

$$\mathfrak{B}_{(7)} := \left\{ x \in \mathscr{L}^1_{\text{loc}}(\mathbb{R}; \mathbb{R}^n) \;\middle|\; \begin{array}{l} Ex \in \mathscr{W}^{1,1}_{\text{loc}}(\mathbb{R}; \mathbb{R}^\ell) \text{ and } x \\ \text{solves (7) for a.a. } t \in \mathbb{R} \end{array} \right\}.$$

Definition 5 (Stabilizability [8, Def. 5.1]) Let $E, A \in \mathbb{R}^{\ell \times n}$. Then $[E, A]$ is called *stabilizable in the behavioral sense* if, and only if,

$$\forall\, x \in \mathfrak{B}_{(7)}\ \exists\, x_0 \in \mathfrak{B}_{(7)} : \ (\forall\, t < 0 : \ x(t) = x_0(t)) \ \wedge \ \lim_{t\to\infty} \text{ess-sup}_{[t,\infty)} \|x_0\| = 0.$$

We are now in a position to derive a duality result.

Lemma 5 (Duality) *Let* $[E, A, B, C] \in \Sigma_{\ell,n,m,p}$. *Then the following statements are equivalent:*

(i) $[E, A, B, C]$ *is stabilizable in the behavioral sense.*
(ii) $[[E, 0], [A, B]]$ *is stabilizable in the behavioral sense.*
(iii) $\left[\begin{bmatrix} E^\top \\ 0 \end{bmatrix}, \begin{bmatrix} A^\top \\ B^\top \end{bmatrix}\right]$ *is stabilizable in the behavioral sense.*
(iv) $[E^\top, A^\top, C^\top, B^\top]$ *is detectable in the behavioral sense.*

Proof It follows from the definition that (i)⇔(ii) and (iii)⇔(iv). By [8, Cor. 5.2], (ii) is equivalent to

$$\forall\, \lambda \in \overline{\mathbb{C}}_+ : \ \text{rk}_{\mathbb{C}}[\lambda E - A, -B] = \text{rk}_{\mathbb{R}(s)}[sE - A, -B].$$

Since ranks are invariant under matrix transpose, we find that (ii) is equivalent to

$$\forall\, \lambda \in \overline{\mathbb{C}}_+ : \ \text{rk}_{\mathbb{C}} \begin{bmatrix} \lambda E^\top - A^\top \\ -B^\top \end{bmatrix} = \text{rk}_{\mathbb{R}(s)} \begin{bmatrix} sE^\top - A^\top \\ -B^\top \end{bmatrix},$$

which, again by [8, Cor. 5.2], is equivalent to (ii). This completes the proof. □

In view of Lemma 5 and [8, Cor. 5.2] we may infer the following.

Corollary 2 (Characterization of stabilizability and detectability) *Let* $[E, A, B, C] \in \Sigma_{\ell,n,m,p}$. *Then the following holds true.*

(i) $[E, A, B, C]$ *is stabilizable in the behavioral sense if, and only if,*

$$\forall\, \lambda \in \overline{\mathbb{C}}_+ : \ \text{rk}_{\mathbb{C}}[\lambda E - A, -B] = \text{rk}_{\mathbb{R}(s)}[sE - A, -B].$$

(ii) $[E, A, B, C]$ *is detectable in the behavioral sense if, and only if,*

$$\forall\, \lambda \in \overline{\mathbb{C}}_+ : \ \text{rk}_{\mathbb{C}} \begin{bmatrix} \lambda E - A \\ -C \end{bmatrix} = \text{rk}_{\mathbb{R}(s)} \begin{bmatrix} sE - A \\ -C \end{bmatrix}.$$

3 Stable Zero Dynamics

In this section we state and prove one of the main results of the present paper and derive some consequences for regular systems.

Theorem 1 (Characterization of stable zero dynamics) *Let* $[E, A, B, C] \in \Sigma_{\ell,n,m,p}$ *be right-invertible and have autonomous zero dynamics. Then the zero dynamics* $\mathscr{ZD}_{(1)}$ *are asymptotically stable if, and only if, the following three conditions hold:*

(i) $[E, A, B, C]$ *is stabilizable in the behavioral sense,*
(ii) $[E, A, B, C]$ *is detectable in the behavioral sense,*
(iii) $[E, A, B, C]$ *has no transmission zeros in* $\overline{\mathbb{C}}_+$.

Proof Since right-invertibility of $[E, A, B, C]$ implies, by Lemma 4, that $\operatorname{rk} C = p$, the assumptions of Lemma 2 are satisfied and we may assume that, without loss of generality, $[E, A, B, C]$ is in the form (3).

⇒: *Step 1*: We show (i). Let

$$T_1(s) := \begin{bmatrix} I_k & 0 & 0 & 0 \\ 0 & I_p & 0 & 0 \\ 0 & 0 & I_{n_3} & 0 \\ -A_{21} & sE_{22} - A_{22} & sE_{23} & -I_m \end{bmatrix} \in \mathbf{Gl}_{n+m}(\mathbb{R}[s])$$

and observe that, since $E_{42} = A_{42} = 0$ by Lemma 4,

$$[sE - A, -B]T_1(s) = \begin{bmatrix} sI_k - Q & -A_{12} & 0 & 0 \\ 0 & 0 & 0 & I_m \\ 0 & sE_{32} & sN - I_{n_3} & 0 \\ 0 & 0 & sE_{43} & 0 \end{bmatrix}.$$

Then, with

$$T_2(s) := \begin{bmatrix} I_k & (sI_k - Q)^{-1}A_{12} & 0 & 0 \\ 0 & I_p & 0 & 0 \\ 0 & 0 & I_{n_3} & 0 \\ 0 & 0 & 0 & -I_m \end{bmatrix} \in \mathbf{Gl}_{n+m}(\mathbb{R}(s)),$$

and

$$T_3(s) := \begin{bmatrix} I_k & 0 & 0 & 0 \\ 0 & I_p & 0 & 0 \\ 0 & -s(sN - I_{n_3})^{-1}E_{32} & I_{n_3} & 0 \\ 0 & 0 & 0 & -I_m \end{bmatrix} \in \mathbf{Gl}_{n+m}(\mathbb{R}[s]),$$

where we note that it follows from (6) that $T_3(s)$ is a polynomial, we obtain

$$[sE - A, -B]T_1(s)T_2(s)T_3(s) = \begin{bmatrix} sI_k - Q & 0 & 0 & 0 \\ 0 & 0 & 0 & I_m \\ 0 & 0 & sN - I_{n_3} & 0 \\ 0 & X(s) & sE_{43} & 0 \end{bmatrix},$$

where $X(s) = -s^2 E_{43}(sN - I_{n_3})^{-1} E_{32} = 0$ by Lemma 4 and (6). Finally,

$$S_1(s) := \begin{bmatrix} I_k & 0 & 0 & 0 \\ 0 & I_p & 0 & 0 \\ 0 & & I_{n_3} & 0 \\ 0 & 0 & -sE_{43}(sN - I_{n_3})^{-1} & -I_m \end{bmatrix} \in \mathbf{Gl}_{n+m}(\mathbb{R}[s])$$

yields

$$S_1(s)[sE - A, -B]T_1(s)T_2(s)T_3(s) = \begin{bmatrix} sI_k - Q & 0 & 0 & 0 \\ 0 & 0 & 0 & I_m \\ 0 & 0 & sN - I_{n_3} & 0 \\ 0 & 0 & 0 & 0 \end{bmatrix}$$

and hence $\mathrm{rk}_{\mathbb{R}(s)}[sE - A, -B] = k + n_3 + m = n + m - p$, since $n_3 = n - k - p$ by Lemma 2. Now let $\lambda \in \overline{\mathbb{C}}_+$ and observe that, by Lemma 3, $\lambda I_k - Q$ is invertible. Hence, the matrices $T_1(\lambda), T_2(\lambda), T_3(\lambda)$ and $S_1(\lambda)$ exist and are invertible. Thus, using the same transformations as above for fixed $\lambda \in \overline{\mathbb{C}}_+$ now, we find that $\mathrm{rk}_{\mathbb{C}}[\lambda E - A, -B] = n + m - p$. This proves (i).

Step 2: We show (ii). Similar to Step 1 it can be shown that

$$\forall\, \lambda \in \overline{\mathbb{C}}_+ : \mathrm{rk}_{\mathbb{C}} \begin{bmatrix} \lambda E - A \\ -C \end{bmatrix} = \mathrm{rk}_{\mathbb{R}(s)} \begin{bmatrix} sE - A \\ -C \end{bmatrix} = n.$$

Step 3: We show (iii). By Corollary 1, the transmission zeros of $[E, A, B, C]$ are the poles of

$$F(s) := A_{21}(sI_k - Q)^{-1} A_{12}.$$

Every pole of $F(s)$ is also an eigenvalue of Q. In view of Lemma 3, we have that $\sigma(Q) \subseteq \mathbb{C}_-$ and so (iii) follows.

$\Leftarrow$: By Lemma 3, we have to show that if $\lambda \in \sigma(Q)$, then $\lambda \in \mathbb{C}_-$. Let $\lambda \in \sigma(Q)$. We distinguish two cases:

Case 1: λ is a pole of $F(s)$. Then, by Corollary 1, λ is a transmission zero of $[E, A, B, C]$ and by (iii) we obtain $\lambda \in \mathbb{C}_-$.

Case 2: λ is not a pole of $F(s)$. Then [7, Lem. 8.3] applied to $[I_k, Q, A_{12}, A_{21}]$ and λ yields that

$$\text{(a) } \operatorname{rk}_{\mathbb{C}}[\lambda I_k - Q, A_{12}] < k \quad \text{or} \quad \text{(b) } \operatorname{rk}_{\mathbb{C}}[\lambda I_k - Q^\top, A_{21}^\top] < k.$$

If (a) holds, then there exists $v_1 \in \mathbb{C}^k \setminus \{0\}$ such that

$$v_1^\top [\lambda I_k - Q, A_{12}] = 0.$$

Let $v_4 \in \mathbb{C}^{(\ell-n)+(p-m)}$ be arbitrary and define

$$v_3^\top := -\lambda v_4^\top E_{43}(\lambda N - I_{n_3})^{-1}.$$

Now observe that

$$(v_1^\top, 0, v_3^\top, v_4^\top) \begin{bmatrix} \lambda I_k - Q & -A_{12} & 0 & 0 \\ -A_{21} & \lambda E_{22} - A_{22} & \lambda E_{23} & I_m \\ 0 & \lambda E_{32} & \lambda N - I_{n_3} & 0 \\ 0 & 0 & \lambda E_{43} & 0 \end{bmatrix} = (0, w^\top, 0, 0),$$

where

$$w^\top = -v_1^\top A_{12} + \lambda v_3^\top E_{32} = -\lambda^2 v_4^\top E_{43}(\lambda N - I_{n_3})^{-1} E_{32} = 0$$

by Lemma 4 and (6). This implies that $\mathscr{K} := \ker[\lambda E - A, -B]^\top \subseteq \mathbb{C}^\ell$ has dimension $\dim \mathscr{K} \geq (\ell - n) + (p - m) + 1$. Therefore,

$$\begin{aligned} \operatorname{rk}_{\mathbb{C}}[\lambda E - A, -B] &\leq \ell - \dim \mathscr{K} \leq n + m - p - 1 \\ &= \operatorname{rk}_{\mathbb{R}(s)}[sE - A, -B] - 1 < \operatorname{rk}_{\mathbb{R}(s)}[sE - A, -B], \end{aligned} \tag{8}$$

where $\operatorname{rk}_{\mathbb{R}(s)}[sE-A,-B] = n+m-p$ has been proved in Step 1 of "$\Rightarrow$". Hence, (8) together with (i) implies that $\lambda \in \mathbb{C}_-$.

If (b) holds, then there exists $v_1 \in \mathbb{C}^k \setminus \{0\}$ such that $v_1^\top [\lambda I_k - Q^\top, A_{21}^\top] = 0$. Therefore,

$$\begin{bmatrix} \lambda I_k - Q & -A_{12} & 0 \\ -A_{21} & \lambda E_{22} - A_{22} & \lambda E_{23} \\ 0 & \lambda E_{32} & \lambda N - I_{n_3} \\ 0 & 0 & \lambda E_{43} \\ 0 & I_p & 0 \end{bmatrix} \begin{pmatrix} v_1 \\ 0 \\ 0 \end{pmatrix} = 0$$

and thus

$$\mathrm{rk}_{\mathbb{C}} \begin{bmatrix} \lambda E - A \\ -C \end{bmatrix} < n = \mathrm{rk}_{\mathbb{R}(s)} \begin{bmatrix} sE - A \\ -C \end{bmatrix}, \tag{9}$$

where $\mathrm{rk}_{\mathbb{R}(s)} \left[\begin{smallmatrix} sE-A \\ -C \end{smallmatrix} \right] = n$ has been proved in Step 2 of "$\Rightarrow$". Hence, (9) together with (ii) implies that $\lambda \in \mathbb{C}_-$. This completes the proof of the theorem. □

Remark 1 It might be surprising that in Step 1 and Step 2 it is calculated that

$$\mathrm{rk}_{\mathbb{R}(s)}[sE - A, -B] = n + m - p \quad \text{and} \quad \mathrm{rk}_{\mathbb{R}(s)} \begin{bmatrix} sE - A \\ -C \end{bmatrix} = n. \tag{10}$$

Because of duality reasons it could be expected that the two ranks satisfy

$$\mathrm{rk}_{\mathbb{R}(s)}[sE-A, -B] = \min\{n+m-p, \ell\} \text{ and } \mathrm{rk}_{\mathbb{R}(s)} \begin{bmatrix} sE - A \\ -C \end{bmatrix} = \min\{\ell+p-m, n\}.$$

Since it is assumed that the zero dynamics are autonomous, it follows that the system

$$\tfrac{\mathrm{d}}{\mathrm{d}t} Ex(t) = Ax(t) + Bu(t), \quad Cx(t) = 0$$

is autonomous and hence no free variables are present. As there are $n + m$ variables and $\ell + p$ equations, it is necessary that $n + m \le \ell + p$. This implies $n + m - p \le \ell$ and $n \le \ell + p - m$, and hence (10).

For regular systems with invertible transfer function we may characterize asymptotic stability of the zero dynamics by Hautus criteria for stabilizability and detectability and the absence of zeros of the transfer function in the closed right complex half-plane (recall Definition 1 for the definition of a zero of a rational matrix function).

Corollary 3 (Regular systems) *Let $[E, A, B, C] \in \Sigma_{n,n,m,m}$ be such that $sE - A$ is regular and $G(s) := C(sE - A)^{-1}B$ is invertible over $\mathbb{R}(s)$. Then the zero dynamics $\mathscr{ZD}_{(1)}$ are asymptotically stable if, and only if, the following three conditions hold:*

(i) $\forall \lambda \in \overline{\mathbb{C}}_+ : \mathrm{rk}_{\mathbb{C}}[\lambda E - A, -B] = n$,

(ii) $\forall \lambda \in \overline{\mathbb{C}}_+ : \mathrm{rk}_{\mathbb{C}} \begin{bmatrix} \lambda E - A \\ -C \end{bmatrix} = n$,

(iii) *$G(s)$ has no zeros in $\overline{\mathbb{C}}_+$.*

Proof Since $G(s) \in \mathbf{Gl}_m(\mathbb{R}(s))$ it follows from Lemma 1 that $\mathscr{ZD}_{(1)}$ are autonomous. Furthermore, $\mathrm{rk}\, C = m$ and hence we may infer from [4, Rem. 4.2.13] that $[E, A, B, C]$ is right-invertible. Now, we may apply Theorem 1 to deduce that $\mathscr{ZD}_{(1)}$ are asymptotically stable if, and only if,

(a) $[E, A, B, C]$ is stabilizable in the behavioral sense,
(b) $[E, A, B, C]$ is detectable in the behavioral sense,
(c) $[E, A, B, C]$ has no transmission zeros in $\overline{\mathbb{C}}_+$.

Since regularity of $sE - A$ gives that $\operatorname{rk}_{\mathbb{R}(s)}[sE - A, -B] = \operatorname{rk}_{\mathbb{R}(s)} \left[\begin{smallmatrix} sE-A \\ -C \end{smallmatrix} \right] = n$, we find that (i)⇔(a) and (ii)⇔(b). (iii)⇔(c) follows from the fact that by [4, Rem. 4.3.3] we have $H(s) = G(s)^{-1}$ for $H(s)$ as in (5) and that transmission zeros of $[E, A, B, C]$ are, by definition, exactly the poles of $H(s)$. □

4 Stabilization

In this section we consider stabilizing control for DAE systems. More precisely, we introduce the concepts of Lyapunov exponent and compatible control and show that for right-invertible systems with autonomous zero dynamics it is possible to assign, via a compatible control, the Lyapunov exponent of the system to a value specified by the zero dynamics.

The usual concept of feedback is the additional application of the relation $u(t) = Fx(t)$ to the system $\frac{\mathrm{d}}{\mathrm{d}t}Ex(t) = Ax(t) + Bu(t)$; for instance, high-gain feedback has been successfully applied to DAEs in [6] in order to achieve stabilization. Feedback can therefore be seen as an additional algebraic constraint that can be resolved for the input. Control in the behavioral sense, or control via interconnection [24], generalizes this approach by also allowing further algebraic relations in which the state not necessarily uniquely determines the input (see also [8, Sec. 5.3]). That is, for given (or to be determined) $K = [K_x, K_u] \in \mathbb{R}^{q \times n} \times \mathbb{R}^{q \times m}$ and $[E, A, B, C] \in \Sigma_{\ell,n,m,p}$ we consider

$$\mathfrak{B}^K_{[E,A,B]} = \left\{ (x, u) \in \mathscr{L}^1_{\mathrm{loc}}(\mathbb{R}; \mathbb{R}^n \times \mathbb{R}^m) \;\middle|\; \begin{array}{l} Ex \in \mathscr{W}^{1,1}_{\mathrm{loc}}(\mathbb{R}; \mathbb{R}^\ell) \text{ and,} \\ \text{for a.a. } t \in \mathbb{R}, \\ \frac{\mathrm{d}}{\mathrm{d}t}Ex(t) = Ax(t) + Bu(t) \\ \qquad\quad 0 = K_x x(t) + K_u u(t) \end{array} \right\}.$$

We call K the *control matrix*, since it induces the control law $K_x x + K_u u \stackrel{\text{a.e.}}{=} 0$. Note that, in principle, one could make the extreme choice $K = I_{n+m}$ to end up with a behavior

$$\mathfrak{B}^K_{[E,A,B]} \subseteq \left\{ (x, u) \in \mathscr{L}^1_{\mathrm{loc}}(\mathbb{R}; \mathbb{R}^n \times \mathbb{R}^m) \;\middle|\; (x, u) \stackrel{\text{a.e.}}{=} 0 \right\},$$

which is obviously asymptotically stable. This, however, is not suitable from a practical point of view. If we assume that the controller is switched on at a certain time $t \in \mathbb{R}$, then this causes a jump from a solution trajectory of the original system $[E, A, B]$ onto a solution within the interconnected behavior $\mathfrak{B}^K_{[E,A,B]}$ (the trivial solution in this case) at time t. Hence, jumps occur in Ex. To avoid this, we use the concept of compatible control.

Definition 6 (Compatible control [8, Def. 5.2]) Let $[E, A, B, C] \in \Sigma_{\ell,n,m,p}$. The control matrix $K = [K_x, K_u] \in \mathbb{R}^{q\times n} \times \mathbb{R}^{q\times m}$ is called *compatible* for $[E, A, B, C]$ if, and only if,

$$\forall\, x^0 \in \left\{\, x^0 \in \mathbb{R}^n \;\middle|\; \exists\,(x,u,y) \in \mathfrak{B}_{(1)} : \; Ex(0) = Ex^0 \,\right\}$$
$$\exists\,(x,u) \in \mathfrak{B}^K_{[E,A,B]} : \; Ex(0) = Ex^0.$$

We construct a compatible control which not only results in an asymptotically stable interconnected system, but also the Lyapunov exponent of the interconnected system is prescribed by the zero dynamics of the nominal system. In order to get a most general definition of the Lyapunov exponent, we use a definition similar to the Bohl exponent in [3, Def. 3.4], not requiring a fundamental solution matrix as in [18].

Definition 7 (Lyapunov exponent) Let $E, A \in \mathbb{R}^{\ell\times n}$. The *Lyapunov exponent* of $[E, A]$ is defined as

$$k_L(E,A) := \inf\left\{\, \mu \in \mathbb{R} \;\middle|\; \begin{array}{r} \exists\, M_\mu > 0\ \forall\, x \in \mathfrak{B}_{(7)} \text{ for a.a. } t \ge s : \\ \|x(t)\| \le M_\mu e^{\mu(t-s)} \|x(s)\| \end{array} \right\}.$$

Note that we use the convention $\inf \emptyset = +\infty$.

The (minimal) exponential decay rate of the (asymptotically stable) zero dynamics of a system can be determined by the Lyapunov exponent of the DAE $\left[\left[\begin{smallmatrix} E & 0 \\ 0 & 0 \end{smallmatrix}\right], \left[\begin{smallmatrix} A & B \\ C & 0 \end{smallmatrix}\right]\right]$.

Lemma 6 (Lyapunov exponent and stable zero dynamics) *Let* $[E, A, B, C] \in \Sigma_{\ell,n,m,p}$ *with autonomous zero dynamics and* $\operatorname{rk} C = p$. *Then, using the notation from Lemma 2 and* k *as in* (4), *we have*

$$\begin{aligned} k_L(\mathcal{ZD}_{(1)}) &:= \inf\left\{\, \mu \in \mathbb{R} \;\middle|\; \begin{array}{r} \exists\, M_\mu > 0\ \forall\, w \in \mathcal{ZD}_{(1)} \text{ for a.a. } t \ge s : \\ \|w(t)\| \le M_\mu e^{\mu(t-s)} \|w(s)\| \end{array} \right\} \\ &= k_L\left(\begin{bmatrix} E & 0 \\ 0 & 0 \end{bmatrix}, \begin{bmatrix} A & B \\ C & 0 \end{bmatrix} \right) \\ &= \begin{cases} \max\{\, \operatorname{Re}\lambda \mid \lambda \in \sigma(Q) \,\}, & \text{if } k > 0 \\ -\infty, & \text{if } k = 0. \end{cases} \end{aligned}$$

Proof The first equality follows from the fact that the trajectories in $\mathcal{ZD}_{(1)}$ can be identified with those in the behavior $\mathfrak{B}_{(7)}$ of the DAE system corresponding to $\left(\left[\begin{smallmatrix} E & 0 \\ 0 & 0 \end{smallmatrix}\right], \left[\begin{smallmatrix} A & B \\ C & 0 \end{smallmatrix}\right]\right)$.

The second equality can be seen by using the decomposition (3): Since the Lyapunov exponent is invariant under transformation of the system (see e.g. [3, Prop. 3.17]) we may assume that, without loss of generality, $[E, A, B, C]$ is in the form (3). Now observe that $(x, u, y) \in \mathscr{Z}\mathscr{D}_{(1)}$, where $x = (x_1, y, x_3)$, if, and only if, $y \overset{\text{a.e.}}{=} 0$, $x_3 \overset{\text{a.e.}}{=} 0$ and $x_1 \in \mathscr{W}^{1,1}_{\text{loc}}(\mathbb{R}; \mathbb{R}^k)$, $u \in \mathscr{L}^1_{\text{loc}}(\mathbb{R}; \mathbb{R}^m)$ satisfy

$$\tfrac{\mathrm{d}}{\mathrm{d}t} x_1 \overset{\text{a.e.}}{=} Q x_1, \quad u \overset{\text{a.e.}}{=} -A_{21} x_1.$$

This equivalence of solution trajectories yields the assertion. □

Note that it follows from Lemmas 3 and 6 that asymptotic stability of the zero dynamics implies exponential stability of the zero dynamics, i.e., any trajectory tends to zero exponentially.

We are now in a position to prove the main result of this section, which states that for right-invertible systems with autonomous zero dynamics there exists a compatible control such that the Lyapunov exponent of the interconnected system is equal to the Lyapunov exponent of the zero dynamics of the nominal system; in particular, this shows that asymptotic stability of the zero dynamics implies that the system can be asymptotically stabilized in the sense that every solution of the interconnected system tends to zero.

Theorem 2 (Compatible and stabilizing control) *Let $[E, A, B, C] \in \Sigma_{\ell,n,m,p}$ be right-invertible with autonomous zero dynamics. If $\dim \mathscr{Z}\mathscr{D}_{(1)} > 0$, then there exists a compatible control matrix $K = [K_x, K_u] \in \mathbb{R}^{q\times n} \times \mathbb{R}^{q\times m}$ for $[E, A, B, C]$ such that*

$$k_L\left(\begin{bmatrix} E & 0 \\ 0 & 0 \end{bmatrix}, \begin{bmatrix} A & B \\ K_x & K_u \end{bmatrix}\right) = k_L(\mathscr{Z}\mathscr{D}_{(1)}). \tag{11}$$

If $\dim \mathscr{Z}\mathscr{D}_{(1)} = 0$, then for all $\mu \in \mathbb{R}$ there exists a compatible control matrix $K = [K_x, K_u] \in \mathbb{R}^{q\times n} \times \mathbb{R}^{q\times m}$ for $[E, A, B, C]$ such that

$$k_L\left(\begin{bmatrix} E & 0 \\ 0 & 0 \end{bmatrix}, \begin{bmatrix} A & B \\ K_x & K_u \end{bmatrix}\right) \le \mu. \tag{12}$$

Proof Since the Lyapunov exponent is invariant under transformation of the system (see e.g. [3, Prop. 3.17]) we may, similar to the proof of Theorem 1, assume that, without loss of generality, $[E, A, B, C]$ is in the form (3). Then, with similar transformations as in Step 1 of the proof of Theorem 1, it can be shown that

$$\forall\, \lambda \in \mathbb{C}: \operatorname{rk}_{\mathbb{C}} \begin{bmatrix} \lambda E_{22} - A_{22} & \lambda E_{23} & I_m \\ \lambda E_{32} & \lambda N - I_{n_3} & 0 \\ 0 & \lambda E_{43} & 0 \end{bmatrix} = \operatorname{rk}_{\mathbb{R}(s)} \begin{bmatrix} sE_{22} - A_{22} & sE_{23} & I_m \\ sE_{32} & sN - I_{n_3} & 0 \\ 0 & sE_{43} & 0 \end{bmatrix},$$

and hence, by [8, Cor. 4.3], the system

$$[\tilde{E}, \tilde{A}, \tilde{B}, \tilde{C}] := \left[\begin{bmatrix} E_{22} & E_{23} \\ E_{32} & N \\ 0 & E_{43} \end{bmatrix}, \begin{bmatrix} A_{22} & 0 \\ 0 & I_{n_3} \\ 0 & 0 \end{bmatrix}, \begin{bmatrix} I_m \\ 0 \\ 0 \end{bmatrix}, [I_p, 0] \right]$$

is controllable in the behavioral sense as in [8, Def. 2.1].

We will now mimic the proof of [8, Thm. 5.4] without repeating all of its arguments: It follows from the above controllability in the behavioral sense and [8, Cor. 3.4] that in the feedback form [8, (3.10)] of $[\tilde{E}, \tilde{A}, \tilde{B}]$ we have $n_{\overline{c}} = 0$. Therefore, for any given $\mu \in \mathbb{R}$ and $\varepsilon > 0$, it is possible to choose F_{11} and K_x in the proof of [8, Thm. 5.4] such that the resulting control matrix $\tilde{K} = [K_1, K_2] \in \mathbb{R}^{q\times(n-k)} \times \mathbb{R}^{q\times m}$ is compatible for $[\tilde{E}, \tilde{A}, \tilde{B}, \tilde{C}]$ and satisfies

$$k_L \left(\begin{bmatrix} \tilde{E} & 0 \\ 0 & 0 \end{bmatrix}, \begin{bmatrix} \tilde{A} & \tilde{B} \\ K_1 & K_2 \end{bmatrix} \right) \leq \mu - \varepsilon. \tag{13}$$

We show that

$$K = [K_x \,|\, K_u] := [K_2 A_{21}, K_1 \,|\, K_2] \in \mathbb{R}^{q\times k} \times \mathbb{R}^{q\times(n-k)} \times \mathbb{R}^{q\times m},$$

is compatible for $[E, A, B, C]$ and satisfies (11) or (12), resp.

Step 1: We show compatibility. Let

$$x^0 \in \left\{ x^0 \in \mathbb{R}^n \;\middle|\; \exists\, (x, u, y) \in \mathfrak{B}_{(1)} : \; Ex(0) = Ex^0 \right\}$$

and partition $x^0 = \left((x_1^0)^\top, (x_2^0)^\top\right)^\top$ with $x_1^0 \in \mathbb{R}^k$, $x_2^0 \in \mathbb{R}^{n-k}$. Then there exist $x_1 \in \mathscr{W}^{1,1}_{\text{loc}}(\mathbb{R}; \mathbb{R}^k)$, $x_2 \in \mathscr{L}^1_{\text{loc}}(\mathbb{R}; \mathbb{R}^{n-k})$ and $u \in \mathscr{L}^1_{\text{loc}}(\mathbb{R}; \mathbb{R}^m)$ such that $\tilde{E}x_2 \in \mathscr{W}^{1,1}_{\text{loc}}(\mathbb{R}; \mathbb{R}^{n-k})$ and

$$\left.\begin{aligned} \tfrac{\mathrm{d}}{\mathrm{d}t} x_1 &\overset{\text{a.e.}}{=} Q x_1 + [A_{12}, 0] x_2, \\ \tfrac{\mathrm{d}}{\mathrm{d}t} \tilde{E} x_2 &\overset{\text{a.e.}}{=} \begin{bmatrix} A_{21} \\ 0 \\ 0 \end{bmatrix} x_1 + \tilde{A} x_2 + \tilde{B} u, \\ x_1(0) &= x_1^0, \\ \tilde{E} x_2(0) &= \tilde{E} x_2^0. \end{aligned}\right\} \tag{14}$$

Therefore,

$$x_2^0 \in \left\{ x_2^0 \in \mathbb{R}^n \;\middle|\; \exists\, (x_2, u, \tilde{C}x_2) \in \mathfrak{B}_{[\tilde{E},\tilde{A},\tilde{B},\tilde{C}]} : \; \tilde{E}x_2(0) = \tilde{E}x_2^0 \right\},$$

where $\mathfrak{B}_{[\tilde{E},\tilde{A},\tilde{B},\tilde{C}]}$ denotes the behavior of (1) corresponding to the system $[\tilde{E}, \tilde{A}, \tilde{B}, \tilde{C}]$, and by compatibility of $[K_1, K_2]$ for $[\tilde{E}, \tilde{A}, \tilde{B}, \tilde{C}]$ there exists

$(x_2, v) \in \mathfrak{B}^{[K_1,K_2]}_{[\tilde{E},\tilde{A},\tilde{B}]}$ such that

$$\begin{aligned} \tfrac{\mathrm{d}}{\mathrm{d}t}\tilde{E}x_2 &\overset{\text{a.e.}}{=} \tilde{A}x_2 + \tilde{B}v, \\ 0 &\overset{\text{a.e.}}{=} K_1x_2 + K_2v, \end{aligned} \tag{15}$$

and $\tilde{E}x_2(0) = \tilde{E}x_2^0$. Define

$$x_1(t) := e^{Qt}x_1^0 + \int_0^t e^{Q(t-s)}[A_{12}, 0]x_2(s)\ \mathrm{d}s\,, \quad t \in \mathbb{R},$$

which is well-defined since $x_2 \in \mathscr{L}^1_{\mathrm{loc}}(\mathbb{R}; \mathbb{R}^{n-k})$, and let $u := v - A_{21}x_1$. Then (x_1, x_2, u) solves (14) and satisfies

$$K_2A_{21}x_1 + K_1x_2 + K_2u \overset{\text{a.e.}}{=} K_2A_{21}x_1 + K_1x_2 + K_2v - K_2A_{21}x_1 \overset{\text{a.e.}}{=} 0,$$

which proves that $[K_2A_{21}, K_1, K_2]$ is compatible for $[E, A, B, C]$.

Step 2: We show that (12) is satisfied in case that $k = 0$ for k as in (4). This follows from (13) since

$$k_L\left(\begin{bmatrix} E & 0 \\ 0 & 0 \end{bmatrix}, \begin{bmatrix} A & B \\ K_x & K_u \end{bmatrix}\right) = k_L\left(\begin{bmatrix} \tilde{E} & 0 \\ 0 & 0 \end{bmatrix}, \begin{bmatrix} \tilde{A} & \tilde{B} \\ K_1 & K_2 \end{bmatrix}\right) \le \mu - \varepsilon$$

with arbitrary $\mu \in \mathbb{R}$ and $\varepsilon > 0$.

Step 3: We show that (11) is satisfied in case that $k > 0$. Denote

$$\mu := k_L(\mathscr{Z}\mathscr{D}_{(1)}) \overset{\text{Lem. 6}}{=} \max\{\, \mathrm{Re}\,\lambda \mid \lambda \in \sigma(Q) \,\}$$

and let $\rho > 0$ be arbitrary. Then there exists $M_\rho > 0$ such that, for all $t \ge 0$, $\|e^{Qt}\| \le M_\rho e^{(\mu+\rho)t}$.

Step 3a: We show "$\ge$" in (11). Since, for any solution $x_1 \in \mathscr{W}^{1,1}_{\mathrm{loc}}(\mathbb{R}; \mathbb{R}^k)$ of $\tfrac{\mathrm{d}}{\mathrm{d}t}x_1 = Qx_1$ we have

$$\big((x_1^\top, 0)^\top, -A_{21}x_1, 0\big) \in \mathfrak{B}^K_{[E,A,B]},$$

it follows that

$$k_L\left(\begin{bmatrix} E & 0 \\ 0 & 0 \end{bmatrix}, \begin{bmatrix} A & B \\ K_x & K_u \end{bmatrix}\right) \ge \mu.$$

Step 3b: We show "$\le$" in (11). Let $(x, u) \in \mathfrak{B}^K_{[E,A,B]}$ and write $x = \left(x_1^\top, x_2^\top\right)^\top$ with $x_1 \in \mathscr{W}^{1,1}_{\text{loc}}(\mathbb{R}; \mathbb{R}^k)$ and $x_2 \in \mathscr{L}^1_{\text{loc}}(\mathbb{R}; \mathbb{R}^{n-k})$. Then we have

$$\begin{aligned} \tfrac{\mathrm{d}}{\mathrm{d}t} x_1 &\overset{\text{a.e.}}{=} Q x_1 + [A_{12}, 0] x_2, \\ \tfrac{\mathrm{d}}{\mathrm{d}t} \tilde{E} x_2 &\overset{\text{a.e.}}{=} \begin{bmatrix} A_{21} \\ 0 \\ 0 \end{bmatrix} x_1 + \tilde{A} x_2 + \tilde{B} u, \\ 0 &\overset{\text{a.e.}}{=} K_2 A_{21} x_1 + K_1 x_2 + K_2 u. \end{aligned}$$

Observe that $(x_2, w := u + A_{21} x_1)$ solves (15) and hence, by (13) for μ and some $\varepsilon > 0$, there exists $M_1 > 0$ such that

$$\text{for a.a. } t \ge s : \quad \left\| \begin{pmatrix} x_2(t) \\ w(t) \end{pmatrix} \right\| \le M_1 e^{(\mu-\varepsilon)(t-s)} \left\| \begin{pmatrix} x_2(s) \\ w(s) \end{pmatrix} \right\|.$$

Therefore,

$$\begin{aligned} \|x_1(t)\| &\le \|e^{Q(t-s)}\| \cdot \|x_1(s)\| + \int_s^t \|e^{Q(t-\tau)}\| \cdot \|[A_{12}, 0]\| \cdot \left\| \begin{pmatrix} x_2(\tau) \\ w(\tau) \end{pmatrix} \right\| \, \mathrm{d}\tau \\ &\le M_\rho e^{(\mu+\rho)(t-s)} \|x_1(s)\| \\ &\quad + M_1 M_\rho e^{(\mu+\rho)(t-s)} \cdot \|[A_{12}, 0]\| \cdot \left\| \begin{pmatrix} x_2(s) \\ w(s) \end{pmatrix} \right\| \underbrace{\int_s^t e^{-(\varepsilon+\rho)(t-\tau)} \, \mathrm{d}\tau}_{\le 1/\varepsilon} \end{aligned}$$

for almost all $t, s \in \mathbb{R}$ with $t \ge s$. This implies that

$$k_L\left(\begin{bmatrix} E & 0 \\ 0 & 0 \end{bmatrix}, \begin{bmatrix} A & B \\ K_x & K_u \end{bmatrix} \right) \le \mu + \rho$$

and since $\rho > 0$ is arbitrary the claim is shown. □

Remark 2 (Construction of the control) The construction of the control K in the proof of Theorem 2 relies on the construction used in [8, Thm. 5.4]. Here we make it precise. We have split up the procedure into several steps.

(i) The first step is to transform the given system $[E, A, B, C] \in \Sigma_{\ell,n,m,p}$ into the form (3). The first transformation which has to be applied in order to achieve this is stated in [4, Thm. 4.1.7] and uses the maximal (E, A, B)-invariant subspace included in $\ker C$. This subspace can be obtained easily via a subspace iteration as described in [4, Lem. 4.1.2]. The second transformation which has to be applied is stated in [4, Thm. 4.2.7]. Denote the resulting system by

$$\begin{bmatrix} s\overline{E}-\overline{A} & \overline{B} \\ \overline{C} & 0 \end{bmatrix} = \begin{bmatrix} P & 0 \\ 0 & I_p \end{bmatrix} \begin{bmatrix} sE-A & B \\ C & 0 \end{bmatrix} \begin{bmatrix} Q & 0 \\ 0 & I_m \end{bmatrix}.$$

(ii) Next we have to consider the subsystem

$$[\tilde{E}, \tilde{A}, \tilde{B}, \tilde{C}] := \left[\begin{bmatrix} E_{22} & E_{23} \\ E_{32} & N \\ 0 & E_{43} \end{bmatrix}, \begin{bmatrix} A_{22} & 0 \\ 0 & I_{n_3} \\ 0 & 0 \end{bmatrix}, \begin{bmatrix} I_m \\ 0 \\ 0 \end{bmatrix}, [I_p, 0] \right]$$

and transform it into a feedback form. To this end we introduce the following notation: For $j \in \mathbb{N}$, we define the matrices

$$N_j = \begin{bmatrix} 0 & & \\ 1 & \ddots & \\ & \ddots & \ddots \\ & & 1 \; 0 \end{bmatrix} \in \mathbb{R}^{j \times j}, \quad K_j = \begin{bmatrix} 1 & 0 & & \\ & \ddots & \ddots & \\ & & 1 & 0 \end{bmatrix}, L_j = \begin{bmatrix} 0 & 1 & & \\ & \ddots & \ddots & \\ & & 0 & 1 \end{bmatrix} \in \mathbb{R}^{(j-1) \times j}.$$

Further, let $e_i^{[j]} \in \mathbb{R}^j$ be the ith canonical unit vector, and, for some multi-index $\alpha = (\alpha_1, \dots, \alpha_l) \in \mathbb{N}^l$, we define

$$\begin{aligned} N_\alpha &= \operatorname{diag}(N_{\alpha_1}, \dots, N_{\alpha_l}) \in \mathbb{R}^{|\alpha| \times |\alpha|}, \\ K_\alpha &= \operatorname{diag}(K_{\alpha_1}, \dots, K_{\alpha_l}) \in \mathbb{R}^{(|\alpha|-l) \times |\alpha|}, \\ L_\alpha &= \operatorname{diag}(L_{\alpha_1}, \dots, L_{\alpha_l}) \in \mathbb{R}^{(|\alpha|-l) \times |\alpha|}, \\ E_\alpha &= \operatorname{diag}(e_{\alpha_1}^{[\alpha_1]}, \dots, e_{\alpha_l}^{[\alpha_l]}) \in \mathbb{R}^{|\alpha| \times l}. \end{aligned}$$

Then it was shown in [19] that a given system can, via state-space, input-space and feedback transformation, be put into a feedback canonical form. Here we use the feedback form from [8, Thm. 3.3], which is not canonical. Since $[\tilde{E}, \tilde{A}, \tilde{B}]$ is controllable in the behavioral sense as in [8, Def. 2.1] and $\operatorname{rk} \tilde{B} = m$, there exist $S \in \mathbf{Gl}_{\ell-k}(\mathbb{R})$, $T \in \mathbf{Gl}_{n-k}(\mathbb{R})$, $V \in \mathbf{Gl}_m(\mathbb{R})$, $F \in \mathbb{R}^{m \times (n-k)}$ such that

$$[s\hat{E} - \hat{A}, \hat{B}] = S[s\tilde{E} - \tilde{A}, \tilde{B}] \begin{bmatrix} T & 0 \\ -F & V \end{bmatrix},$$

where

$$[\hat{E}, \hat{A}, \hat{B}] = \left[\begin{bmatrix} I_{|\alpha|} & 0 & 0 & 0 & 0 \\ 0 & K_\beta & 0 & 0 & 0 \\ 0 & 0 & L_\gamma^\top & 0 & 0 \\ 0 & 0 & 0 & K_\delta^\top & 0 \\ 0 & 0 & 0 & 0 & N_\kappa \end{bmatrix}, \begin{bmatrix} N_\alpha^\top & 0 & 0 & 0 & 0 \\ 0 & L_\beta & 0 & 0 & 0 \\ 0 & 0 & K_\gamma^\top & 0 & 0 \\ 0 & 0 & 0 & L_\delta^\top & 0 \\ 0 & 0 & 0 & 0 & I_{|\kappa|} \end{bmatrix}, \begin{bmatrix} E_\alpha & 0 \\ 0 & 0 \\ 0 & E_\gamma \\ 0 & 0 \\ 0 & 0 \end{bmatrix} \right],$$

for some multi-indices $\alpha, \beta, \gamma, \delta, \kappa$.

(iii) Let $\mu \in \mathbb{R}$ be arbitrary. We construct a compatible control in the behavioral sense for $[\hat{E}, \hat{A}, \hat{B}]$ such that the interconnected system has Lyapunov exponent smaller or equal to μ. Let $F_{11} \in \mathbb{R}^{L(\alpha)\times|\alpha|}$ be such that

$$\max\{\operatorname{Re}\lambda \mid \lambda \in \sigma(N_\alpha + E_\alpha F_{11})\} \le \mu.$$

This can be achieved as follows: For $j = 1, \ldots, L(\alpha)$, consider vectors

$$a_j = -[a_{j\alpha_j-1}, \ldots, a_{j0}] \in \mathbb{R}^{1\times\alpha_j}.$$

Then, for

$$F_{11} = \operatorname{diag}(a_1, \ldots, a_{L(\alpha)}) \in \mathbb{R}^{L(\alpha)\times|\alpha|},$$

the matrix $N_\alpha + E_\alpha F_{11}$ is diagonally composed of companion matrices, whence, for

$$p_j(s) = s^{\alpha_j} + a_{j\alpha_j-1}s^{\alpha_j-1} + \ldots + a_{j0} \in \mathbb{R}[s]$$

the characteristic polynomial of $N_\alpha + E_\alpha F_{11}$ is given by

$$\det(sI_{|\alpha|} - (N_\alpha + E_\alpha F_{11})) = \prod_{j=1}^{L(\alpha)} p_j(s).$$

Hence, choosing the coefficients a_{ji}, $j = 1, \ldots, L(\alpha)$, $i = 0, \ldots, \alpha_j$ such that the roots of the polynomials $p_1(s), \ldots, p_{L(\alpha)}(s) \in \mathbb{R}[s]$ are all smaller or equal to μ yields the assertion.

Now we find that

$$k_L\left(\begin{bmatrix} I_{|\alpha|} & 0 \\ 0 & 0 \end{bmatrix}, \begin{bmatrix} N_\alpha & E_\alpha \\ F_{11} & -I_{L(\alpha)} \end{bmatrix}\right) \le \mu.$$

Furthermore, by the same reasoning as above, for

$$a_j = [a_{j\beta_j-2}, \ldots, a_{j0}, 1] \in \mathbb{R}^{1\times\beta_j}$$

with the property that the roots of the polynomials

$$p_j(s) = s^{\beta_j} + a_{j\beta_j-1}s^{\beta_j-1} + \ldots + a_{j0} \in \mathbb{R}[s]$$

are all smaller or equal to μ for $j = 1, \ldots, L(\alpha)$, the choice

$$K_x = \operatorname{diag}(a_1, \ldots, a_{L(\beta)}) \in \mathbb{R}^{L(\beta)\times|\beta|}$$

leads to

$$k_L\left(\begin{bmatrix} K_\beta \\ 0 \end{bmatrix}, \begin{bmatrix} L_\beta \\ K_x \end{bmatrix}\right) \le \mu.$$

Therefore, the control matrix

$$\hat{K} = [\hat{K}_1, \hat{K}_2] = \begin{bmatrix} F_{11} & 0 & 0\,0\,0 & -I_{L(\alpha)} & 0 \\ 0 & K_x & 0\,0\,0 & 0 & 0 \end{bmatrix} \in \mathbb{R}^{q\times(n-k)} \times \mathbb{R}^{q\times m},$$

where $q = L(\alpha) + L(\beta)$, establishes that

$$k_L\left(\begin{bmatrix} \hat{E} & 0 \\ 0 & 0 \end{bmatrix}, \begin{bmatrix} \hat{A} & \hat{B} \\ \hat{K}_1 & \hat{K}_2 \end{bmatrix}\right) \le \mu.$$

Since the differential variables can be arbitrarily initialized in any of the previously discussed subsystems, the constructed control $\hat{K}$ is also compatible for $[\hat{E}, \hat{A}, \hat{B}]$.

(iv) We show that $\hat{K}$ leads to a compatible control $\tilde{K}$ for $[\tilde{E}, \tilde{A}, \tilde{B}]$ such that the interconnected system has Lyapunov exponent smaller or equal to μ. Observe that

$$\begin{bmatrix} S^{-1} & 0 \\ 0 & I_q \end{bmatrix} \begin{bmatrix} s\hat{E} - \hat{A} & \hat{B} \\ \hat{K}_1 & \hat{K}_2 \end{bmatrix} \begin{bmatrix} T^{-1} & 0 \\ V^{-1}FT^{-1} & V^{-1} \end{bmatrix} = \begin{bmatrix} s\tilde{E} - \tilde{A} & \tilde{B} \\ \hat{K}_1 + \hat{K}_2 V^{-1} F T^1 & \hat{K}_2 V^{-1} \end{bmatrix}$$

and hence, by invariance of the Lyapunov exponent under transformation of the system (see e.g. [3, Prop. 3.17]), we find that for

$$[K_1, K_2] := [\hat{K}_1 + \hat{K}_2 V^{-1} F T^1, \hat{K}_2 V^{-1}] \in \mathbb{R}^{q\times(n-k)} \times \mathbb{R}^{q\times m},$$

we have

$$k_L\left(\begin{bmatrix} \tilde{E} & 0 \\ 0 & 0 \end{bmatrix}, \begin{bmatrix} \tilde{A} & \tilde{B} \\ K_1 & K_2 \end{bmatrix}\right) \le \mu.$$

(v) If $k = \dim \mathcal{ZD}_{(1)} = 0$, then we can choose $\mu \in \mathbb{R}$ as we like and obtain

$$k_L\left(\begin{bmatrix} \overline{E} & 0 \\ 0 & 0 \end{bmatrix}, \begin{bmatrix} \overline{A} & \overline{B} \\ \overline{K}_x := K_1 & \overline{K}_u := K_2 \end{bmatrix}\right) = k_L\left(\begin{bmatrix} \tilde{E} & 0 \\ 0 & 0 \end{bmatrix}, \begin{bmatrix} \tilde{A} & \tilde{B} \\ K_1 & K_2 \end{bmatrix}\right) \le \mu.$$

If $k > 0$, then we can choose $\mu < k_L(\mathcal{ZD}_{(1)})$ and obtain, with

$$[\overline{K}_x \,|\, \overline{K}_u] := [K_2 A_{21}, K_1 \,|\, K_2] \in \mathbb{R}^{q\times k} \times \mathbb{R}^{q\times(n-k)} \times \mathbb{R}^{q\times m},$$

that

$$k_L\left(\begin{bmatrix}\overline{E} & 0\\ 0 & 0\end{bmatrix}, \begin{bmatrix}\overline{A} & \overline{B}\\ \overline{K}_x & \overline{K}_u\end{bmatrix}\right) = k_L(\mathscr{X}\mathscr{D}_{(1)}).$$

This is shown in the proof of Theorem 2.

(vi) The desired compatible control K for $[E, A, B, C]$ is now given by

$$K = [\overline{K}_x Q^{-1}, \overline{K}_u].$$

Note that the practical computation of the decompositions in (i) and (ii) is in general not numerically stable. This can be achieved by using orthogonal transformations and condensed forms as in [11]. It seems that with some effort the form (3) in (i) can also be obtained with orthogonal transformations, but this needs to be investigated in detail. Instead of the feedback form from [8, Thm. 3.3] in (ii) a condensed form from [11] could be used. However, in the present work we do not focus on the numerical aspect.

Remark 3 (Implementation of the control) As explained in the beginning of this section, the control law

$$K_x x(t) + K_u u(t) = 0$$

cannot necessarily be solved for $u(t)$. This raises the question for the implementation of the controller. There are basically two perspectives in this regard:

(i) In order to implement the controller it is necessary that all free variables of the open-loop system $\frac{\mathrm{d}}{\mathrm{d}t}Ex(t) = Ax(t) + Bu(t)$ can be manipulated. The free variables of the system can be identified via the quasi-Kronecker form [9, 10] of the pencil $s[E, 0] - [A, B]$; each underdetermined block $sK_{\beta_i} - L_{\beta_i}$ in the quasi-Kronecker form yields one free variable, i.e., there are $L(\beta)$ free variables in the system. The set of free variables may consist of input variables as well as state variables and not necessarily all input variables are free variables. For the implementation of the control, the free variables are treated as controls and the control law can be solved for the free variables. A similar approach has been discussed in [12].

(ii) For an alternative approach, where we do not wish to reinterpret variables, we use the fact that (cf. also Remark 2 (iii)) the control law can be rewritten in the form

$$\begin{bmatrix}K_1\\ K_2\end{bmatrix} x(t) + \begin{bmatrix}I_r & 0\\ 0 & 0\end{bmatrix}\begin{pmatrix}u_1(t)\\ u_2(t)\end{pmatrix} = 0,$$

where a suitable input space transformation has been performed. Then we may solve the first row for $u_1(t)$ and implement this control. It only remains to implement the algebraic condition $K_2 x(t) = 0$. In practice, this relation can be implemented by integrating appropriate components (such as dampers or resistors) into the given plant. In particular, it is not necessary to (actively) manipulate specific state variables, only the implementation of an algebraic relation between some of the state variables is necessary.

Theorem 2 shows that right-invertible systems $[E, A, B, C] \in \Sigma_{\ell,n,m,p}$ with asymptotically stable zero dynamics can be stabilized by a compatible control so that any solution of the interconnected system tends to zero. It is well known [15, Rem. 6.1.3] that any linear ODE system with asymptotically stable zero dynamics (and $p = m$) is stabilizable by *state feedback*, i.e., the compatible control is of the form $u = Fx$. While there is a lot of literature on the state feedback stabilization of linear DAEs, see e.g. [8, 17, 23, 25], it seems that the stabilization problem for systems with asymptotically stable zero dynamics has not been investigated yet. For regular DAE systems we obtain the following result.

Proposition 1 (Regular systems and state feedback) *If $sE - A \in \mathbb{R}[s]^{n\times n}$ in Theorem 2 is regular, then the compatible control K can be chosen as a state feedback in each case, i.e., $K = [K_x, -I_m] \in \mathbb{R}^{m\times n} \times \mathbb{R}^{m\times m}$.*

Proof We use the procedure presented in Remark 2 and modify it at some instances.

(i) Consider the system $[\hat{E}, \hat{A}, \hat{B}, \hat{C}]$ from Remark 2 (ii) and observe that, using the same argument as in [8, Rem. 5.3 (i)], we obtain $L(\delta) = 0$ and $L(\beta) = L(\gamma)$.
(ii) For any multi-index $\eta \in \mathbb{N}^l$ let

$$F_\eta := \operatorname{diag}(e_1^{[\eta_1]}, \ldots, e_1^{[\eta_l]}) \in \mathbb{R}^{|\eta|\times l}.$$

A straightforward calculation yields that there exists a permutation matrix $P \in \mathbb{R}^{\xi\times\xi}$, $\xi := |\beta| + |\gamma| - L(\gamma)$, such that

$$P\left(s\begin{bmatrix} K_\beta & 0 \\ 0 & L_\gamma^\top \end{bmatrix} - \begin{bmatrix} L_\beta & 0 \\ E_\gamma F_\beta^\top & K_\gamma^\top \end{bmatrix}\right) = s\tilde{N} - I_\xi,$$

where $\tilde{N} \in \mathbb{R}^{\xi\times\xi}$ is nilpotent.
(iii) Changing the control matrix $\hat{K}$ in Remark 2 (iii) to

$$\hat{K} = [\hat{K}_1, \hat{K}_2] = \begin{bmatrix} F_{11} & 0 & 0\,0\,0 & -I_{L(\alpha)} & 0 \\ 0 & F_\beta^\top & 0\,0\,0 & 0 & -I_{L(\gamma)} \end{bmatrix} \in \mathbb{R}^{m\times(n-k)} \times \mathbb{R}^{m\times m},$$

where it is worth noting that $L(\alpha) + L(\gamma) = m$, and invoking the observation in (ii), we obtain the same result for the Lyapunov exponent, and the control can be equivalently expressed as a state feedback

$$u_1 = F_{11}x_1, \quad u_2 = F_\beta^\top x_2.$$

Since $\hat{K}_2 = -I_m$ we can write $[K_1, K_2]$ in Remark 2 (iv) as

$$[K_1, K_2] = [V\hat{K}_1 - FT^{-1}, -I_m]$$

and, furthermore, we have $\overline{K}_u = -I_m$ in Remark 2 (vi). Therefore, the compatible control K is a state feedback. □

Acknowledgements I am indebted to Achim Ilchmann (Ilmenau University of Technology) for several constructive discussions.

References

1. Adams, R.A.: Sobolev Spaces. No. 65 in Pure and Applied Mathematics. Academic, New York/London (1975)
2. Bender, D., Laub, A.: The linear quadratic optimal regulator problem for descriptor systems. IEEE Trans. Autom. Control **32**, 672–688 (1987)
3. Berger, T.: Bohl exponent for time-varying linear differential-algebraic equations. Int. J. Control **85**(10), 1433–1451 (2012)
4. Berger, T.: On differential-algebraic control systems. Ph.D. thesis, Institut für Mathematik, Technische Universität Ilmenau, Universitätsverlag Ilmenau, Ilmenau (2013)
5. Berger, T.: Zero dynamics and funnel control of general linear differential-algebraic systems (2013). Submitted for publication. Preprint available from the website of the author
6. Berger, T., Ilchmann, A., Reis, T.: Normal forms, high-gain, and funnel control for linear differential-algebraic systems. In: Biegler, L.T., Campbell, S.L., Mehrmann, V. (eds.) Control and Optimization with Differential-Algebraic Constraints. Advances in Design and Control, vol. 23, pp. 127–164. SIAM, Philadelphia (2012)
7. Berger, T., Ilchmann, A., Reis, T.: Zero dynamics and funnel control of linear differential-algebraic systems. Math. Control Signals Syst. **24**(3), 219–263 (2012)
8. Berger, T., Reis, T.: Controllability of linear differential-algebraic systems – a survey. In: Ilchmann, A., Reis, T. (eds.) Surveys in Differential-Algebraic Equations I. Differential-Algebraic Equations Forum, pp. 1–61. Springer, Berlin/Heidelberg (2013)
9. Berger, T., Trenn, S.: The quasi-Kronecker form for matrix pencils. SIAM J. Matrix Anal. Appl. **33**(2), 336–368 (2012)
10. Berger, T., Trenn, S.: Addition to "The quasi-Kronecker form for matrix pencils". SIAM J. Matrix Anal. Appl. **34**(1), 94–101 (2013)
11. Bunse-Gerstner, A., Byers, R., Mehrmann, V., Nichols, N.K.: Feedback design for regularizing descriptor systems. Linear Algebra Appl. **299**, 119–151 (1999)
12. Campbell, S.L., Kunkel, P., Mehrmann, V.: Regularization of linear and nonlinear descriptor systems. In: Biegler, L.T., Campbell, S.L., Mehrmann, V. (eds.) Control and Optimization with Differential-Algebraic Constraints. Advances in Design and Control, vol. 23, pp. 17–36. SIAM, Philadelphia (2012)
13. Francis, B.A., Wonham, W.M.: The role of transmission zeros in linear multivariable regulators. Int. J. Control **22**(5), 657–681 (1975)
14. Hou, M., Müller, P.C.: Observer design for descriptor systems. IEEE Trans. Autom. Control **44**(1), 164–168 (1999)
15. Isidori, A.: Nonlinear Control Systems. Communications and Control Engineering Series, 3rd edn. Springer, Berlin (1995)
16. Kailath, T.: Linear Systems. Prentice-Hall, Englewood Cliffs (1980)

17. Lewis, F.L.: A survey of linear singular systems. IEEE Proc. Circuits Syst. Signal Process. **5**(1), 3–36 (1986)
18. Linh, V.H., Mehrmann, V.: Lyapunov, Bohl and Sacker-Sell spectral intervals for differential-algebraic equations. J. Dyn. Differ. Equ. **21**, 153–194 (2009)
19. Loiseau, J.J., Özçaldiran, K., Malabre, M., Karcanias, N.: Feedback canonical forms of singular systems. Kybernetika **27**(4), 289–305 (1991)
20. Polderman, J.W., Willems, J.C.: Introduction to Mathematical Systems Theory. A Behavioral Approach. Springer, New York (1998)
21. Rosenbrock, H.H.: State Space and Multivariable Theory. Wiley, New York (1970)
22. Trentelman, H.L., Stoorvogel, A.A., Hautus, M.L.J.: Control Theory for Linear Systems. Communications and Control Engineering. Springer, London (2001)
23. Varga, A.: On stabilization methods of descriptor systems. Syst. Control Lett. **24**, 133–138 (1995)
24. Willems, J.C.: The behavioral approach to open and interconnected systems. IEEE Control Syst. Mag. **27**(6), 46–99 (2007)
25. Xu, S., Van Dooren, P., Ştefan, R., Lam, J.: Robust stability and stabilization for singular systems with state delay and parameter uncertainty. IEEE Trans. Autom. Control **47**(7), 1122–1128 (2002)

Design of Causal Observers for Nonlinear Descriptor Systems

Daniel Labisch and Ulrich Konigorski

Abstract This contribution not only provides a new necessary and sufficient condition for causal observability of nonlinear descriptor systems but also a method to design the causal observer. The approach is based on the transformation of the descriptor system into a state-space form, the so called coupled state-space system. This description exists for all regular descriptor systems no matter if they are proper or not. If the new condition is satisfied, the coupled state-space system can be modified and can be used to design a state-space observer. It is shown, that this observer is also a causal observer for the original descriptor system. Two examples illustrate the capability of the new approach.

Keywords Observer design • Nonlinear descriptor system

Mathematics Subject Classification (2010) 93B07 · 93B17

1 Introduction

Designing an observer for a state-space system directly leads to an observer in state-space form. Designing an observer for descriptor systems, e.g. the method in [8] provides an observer in descriptor form. To reduce the effort for the numerical simulation of the observer in the control unit an observer in state-space form for the descriptor system is preferred. This is called a *causal observer*. Among others in [12] and [13] a design procedure for linear and impulse observable descriptor systems is presented. Less restrictive are the sufficient conditions in [18] and [31]

D. Labisch (✉)
Siemens AG, Industry Sector, Industry Automation Division, Karlsruhe, Germany
e-mail: daniel.labisch@siemens.com

U. Konigorski
Department of Control Engineering and Mechatronics, Technische Universität Darmstadt, Darmstadt, Germany
e-mail: ukonigorski@iat.tu-darmstadt.de

S. Schöps et al. (eds.), *Progress in Differential-Algebraic Equations*, Differential-Algebraic Equations Forum, DOI 10.1007/978-3-662-44926-4_3

which require the detectability of the descriptor system together with another rank assumption. Firstly in [1] it was shown that it is possible to design a causal observer for linear descriptor systems if and only if the descriptor system is detectable and the fast controllable subsystem is observable. However, the costly transformation in *Weierstrass-Kronecker* form is needed. Later in [19, 20] necessary and sufficient conditions were given without computing the Weierstrass-Kronecker form. With the same assumptions a linear observer in descriptor form is transformed based on parametric eigenstructure assignment into an equivalent observer in state-space form in [28] and [29].

Boutayeb and Darouach [6] use linear methods to design an observer for nonlinear descriptor systems linearized at an equilibrium point. Thereby the impulse observability is needed whereas [2] demands an impulse-free linearized descriptor system. Many publications deal with nonlinear descriptor systems which can be described by a linear one disturbed by some nonlinearity (see e.g. [10, 11, 15, 22, 23, 27, 33, 35, 36, 38, 39]). If the nonlinearity satisfies a local Lipschitz condition, it is compensated and the design is carried on with linear methods. Besides additional requirements as impulse observability or even observability of the fast subsystem this class of descriptor systems is very restrictive. In [4] as well as in [14] these limitations are not required, but only observers for index-1 descriptor systems are designed. With the observability of implicit descriptor systems [37] is concerned. Additionally, in [17] the influence of unknown parameters is considered. But both give no design procedures for observers. In contrast, [40] transfers the descriptor observer design to a state-space observer design where the descriptor system has no inputs. Trouble with differentiating behavior is completely avoided by this procedure. Limited to a class of mechanical descriptor systems causal observers are designed directly in [21].

In this contribution a completely new approach [24] to design causal observers is introduced. The descriptor system is transformed into a state-space system which has the same solutions on a submanifold of the state manifold. Rendering this submanifold invariant and attractive gives the state-space system its name *coupled state-space system*. This system not only exists for proper descriptor systems but for all regular descriptor systems. The coupled state-space system will be modified, such that it can be used for a state-space observer design. It will be shown that this observer, if it exists, is always an observer for the descriptor system, hence, a causal observer.

During the design procedure a condition for the success of the approach is presented which becomes a new necessary and sufficient condition for causal observability of observable descriptor systems. Additional assumptions for the descriptor system, such as impulse observability or properness, are not needed in this approach.

The contribution is organized as follows. Section 2 introduces the description and the key properties of descriptor systems. The derivation of the coupled state-space system is carried out in Sect. 3. Also important properties of this system description are discussed there. Section 4 is focused on the causal observer design and derives the new condition for causal observability. We exemplify the observer design with two models in Sect. 5. Firstly, a linear descriptor system of a three-mass oscillator shows the main ideas of the approach. Secondly, an academic nonlinear example illustrates more complex issues for nonlinear descriptor systems.

2 Properties of Descriptor Systems

We focus on descriptor systems in semi-explicit form [16]

$$\dot{x}_1 = f(x_1, x_2, u) \tag{1a}$$

$$0 = g(x_1, x_2, u) \tag{1b}$$

$$y = h(x_1, x_2, u)\,. \tag{1c}$$

$x_1 \in \mathscr{M}_{\mathrm{x}_1}$ describes the *differential variable* and $x_2 \in \mathbb{R}^m$ the *algebraic variable* of the system. Together they constitute the *descriptor* $x = (x_1, x_2) \in \mathscr{M}_{\mathrm{x}}$. $\mathscr{M}_{\mathrm{x}_1}$ and $\mathscr{M}_{\mathrm{x}}$ are smooth r- and n-dimensional manifolds with $n = r + m$. The input is denoted by $u \in \mathbb{R}^p$ and the output by $y \in \mathbb{R}^q$. The vector field $f : (\mathscr{M}_{\mathrm{x}_1} \times \mathbb{R}^m \times \mathbb{R}^p) \to T_{x_1}\mathscr{M}_{\mathrm{x}_1}$ and the maps $g : (\mathscr{M}_{\mathrm{x}_1} \times \mathbb{R}^m \times \mathbb{R}^p) \to \mathbb{R}^m$ and $h : (\mathscr{M}_{\mathrm{x}_1} \times \mathbb{R}^m \times \mathbb{R}^p) \to \mathbb{R}^q$ are assumed to be sufficiently smooth.

The crucial property of descriptor systems is regularity. As it is associated with the solution of the system, we have to define solutions first.

Definition 1 ([30]) A function $x : \mathbb{I} \to \mathscr{M}_{\mathrm{x}}$ is called a *solution* of the system (1) if $x \in \mathscr{C}^1(\mathbb{I}, \mathscr{M}_{\mathrm{x}})$ and x satisfies (1) pointwise for a given input function u. It is called a solution of the initial value problem consisting of (1) and

$$x(t_0) = \hat{x} \tag{2}$$

if x is a solution of (1) and satisfies (2). An initial value (2) is called *consistent* if the corresponding initial value problem has at least one solution.

Definition 2 ([30]) A descriptor system (1) is called *regular* in a neighborhood of $(\hat{x}, \hat{u})$ satisfying (1) if it has a unique solution for every sufficiently smooth input function u in a neighborhood of $\hat{u}$ and every initial value in a neighborhood of $\hat{x}$ that is consistent for the system.

To get solutions $x \in \mathscr{C}^1(\mathbb{I}, \mathscr{M}_{\mathrm{x}})$ sufficiently smooth input functions are required. Less restrictive assumptions can be considered in general but they will not lead to differentiable solutions.

Beside regularity another important property is properness which is defined next.

Definition 3 A regular descriptor system (1) is called *proper* if it has a unique solution for every input function $u \in \mathscr{C}^1(\mathbb{I}, U \subset \mathbb{R}^p)$ in a neighborhood of $\hat{u}$ and every initial value in a neighborhood of $\hat{x}$ that is consistent for the system. If this is not the case, the descriptor system is called *improper*. If the unique solution already exists for every input function $u \in \mathscr{C}^0(\mathbb{I}, U \subset \mathbb{R}^p)$, the descriptor system is called *strictly proper*.

Hence, a descriptor system is proper if the solution does not depend on derivatives of the input function. Regularity and properness are local properties which can only be checked in local neighborhoods.

3 Coupled State-Space Representation

Ignoring the algebraic equation of (1) leads to the fictitious state-space system

$$\dot{x}_1 = f(x_1, x_2, u) \tag{3a}$$

$$z = g(x_1, x_2, u) \tag{3b}$$

$$y = h(x_1, x_2, u) \tag{3c}$$

consisting of the r-dimensional state $x_1 \in \mathcal{M}_{x_1}$. $x_2 \in \mathbb{R}^m$ is taken as an additional input and $z \in \mathbb{R}^m$ as an additional output. This state-space system is closely related to the descriptor system and is suitable to check for regularity as the next lemma shows.

Lemma 1 ([26]) *Let* $(\hat{x}_1, \hat{x}_2)$ *be a consistent initial value for the descriptor system* (1). *The descriptor system is regular in a neighborhood of* $(\hat{x}_1, \hat{x}_2, \hat{u})$ *if and only if* (3) *is invertible in a neighborhood of* $(\hat{x}_1, \hat{x}_2, \hat{u})$ *with respect to the input* x_2 *and output* z *(which will be denoted by* $\langle z, x_2 \rangle$*).*

Proof $\Leftarrow$: Let $x(t) = (x_1(t), x_2(t))$ be a function which pointwise satisfies (1) together with the input function $u(t)$. Then $x(t)$ and $u(t)$ pointwise satisfy (3) with $z(t) = 0$. As (3) is invertible with respect to $\langle z, x_2 \rangle$, $x_2(t)$ is uniquely determined by $z(t) = 0$ and $\hat{x} = x(t_0)$. Hence, for a given $u(t)$ there exists a unique solution satisfying (3) with $z(t) = 0$. As there exists a unique solution of (1), (1) is regular.

$\Rightarrow$: Let the descriptor system (1) be regular. Then the solution $x(t) = (x_1(t), x_2(t))$ pointwise satisfying (1) for a given $u(t)$ and a consistent initial value $\hat{x} = x(t_0)$ is unique. Suppose (3) is not invertible with respect to $\langle z, x_2 \rangle$. Then a $\tilde{x}(t) = (\tilde{x}_1(t), \tilde{x}_2(t))$ exists for which $\tilde{x}_2(t) \neq x_2(t)$, $\tilde{x}_2(t_0) = x_2(t_0)$ and $\tilde{x}_1(t_0) = x_1(t_0)$ holds. Hence, $\tilde{x}(t)$ pointwise satisfies (3) with $z(t) = 0$ for the input function $u(t)$. Therefore, $\tilde{x}(t)$ pointwise satisfies (1). As (1) posses two solutions for the same initial value and input function, the descriptor system is not regular. This contradicts the assumption. □

The following algorithm will start with (3) and create a coupled state-space system. Like index-reduction methods the basic idea is to differentiate the algebraic equations with respect to time to create new constraints. The constraints are differentiated until the algebraic variables appear and the resulting system of equations can be uniquely solved for the algebraic variables. Interpreting this as a fictitious controller for (3), a new controlled system is created. All equations not depending on x_2 define a submanifold $\mathcal{N}$ of the state manifold $\mathcal{M}_{x_1}$ which contains all solutions of the descriptor system. As this submanifold is invariant for the controlled system, solutions for initial values on $\mathcal{N}$ stay in $\mathcal{N}$ and agree with the solution of the descriptor system. For numerical reasons the submanifold $\mathcal{N}$ is additionally rendered attractive.

If the descriptor system is improper, derivatives of the input will appear during the algorithm. To avoid this, the descriptor system is extended by integrators at the inputs. If there is more than one input acting on a constraint, an input transformation will be carried out to ensure that only one integrator is necessary to continue the algorithm.

For a better readability the algorithm is carried out for the first algebraic equation at the beginning (Steps 1–3). Afterwards, the remaining algebraic equations are considered (Steps 4–8) in which Steps 4–7 equal Steps 1–3.

Step 0 The system (3) is denoted by Σ_0 and $\tilde{x}_1 := x_1$, $\tilde{u} := u$, $\tilde{f}(\tilde{x}_1, \tilde{u}, x_2) := f(x_1, x_2, u)$, $\gamma_0(\tilde{x}_1, \tilde{u}, x_2) := g(x_1, x_2, u)$. The m functions in γ_0 are denoted by $\gamma_0^1, \ldots, \gamma_0^m$. Let $\mathcal{M}_0 := \mathcal{M}_1 := \mathcal{M}_{x_1}$ be the state manifold for Σ_0. κ_j counts the numbers of derivatives of the jth algebraic equation and μ the order of the dynamic extension which is necessary. In the beginning, $\kappa_j = 0$ $j \in \underline{m}$ and $\mu = 0$ are set.

Step 1 If $\partial_{x_2}\gamma_{\kappa_1}^1 \neq 0$ holds in a neighborhood of $\hat{\tilde{x}}_1$ on $\mathcal{M}_1$, we directly go to step 3 as there are no additional derivatives of the equation necessary.
If $\partial_u \gamma_{\kappa_1}^1 \neq 0$ holds in a neighborhood of $\hat{\tilde{x}}_1$ on $\mathcal{M}_1$, μ is increased by 1 and the system is dynamically extended by an integrator. Therefore, a regular input transformation $\tilde{u} = \beta_\mu(\tilde{x}_1, \tilde{v})$, $\tilde{v} \in \mathbb{R}^p$ is needed which locally satisfies $\operatorname{rank} \partial_{\tilde{v}}\beta_\mu = p$. The transformation is selected such that with the partitioning $\tilde{v} = (\tilde{v}_1, \tilde{v}_2)$, $\tilde{v}_1 \in \mathbb{R}^{p-1}$, $\tilde{v}_2 \in \mathbb{R}$ only $\tilde{v}_2$ has influence on $\gamma_{\kappa_1}^1$. Hence, locally $\partial_{\tilde{v}_1}(\gamma_{\kappa_1}^1 \circ \beta_\mu) = 0$ and $\partial_{\tilde{v}_2}(\gamma_{\kappa_1}^1 \circ \beta_\mu) \neq 0$ hold. In addition, β_μ is to be chosen such that at $\hat{\tilde{x}}_1$ $\partial_{x_1}\tilde{f} = (\partial_{\tilde{x}_1}\tilde{f} \circ \beta_\mu) \circ \beta_\mu^{-1}$, $\partial_{x_1}\gamma_{\kappa_1}^1 = (\partial_{\tilde{x}_1}\gamma_{\kappa_1}^1 \circ \beta_\mu) \circ \beta_\mu^{-1}$, and $\partial_{x_1}\gamma_0^i = (\partial_{\tilde{x}_1}\gamma_0^i \circ \beta_\mu) \circ \beta_\mu^{-1}$ for $1 < i \leq m$ is valid. Extending the state with $\tilde{v}_2$ leads to $\check{x}_1 := (\tilde{x}_1, \tilde{v}_2)$ and $\check{u} := (\check{u}_1, \check{u}_2)$ with $\check{u}_1 = \tilde{v}_1$. With $\dot{\tilde{v}}_2 = \check{u}_2$ it follows

$$\check{f}(\check{x}_1, \check{u}, x_2) := \begin{bmatrix} \tilde{f}(\tilde{x}_1, \beta_\mu(\tilde{x}_1, \tilde{v}), x_2) \\ \check{u}_2 \end{bmatrix}. \tag{4}$$

Moreover, $\mathcal{M}_0$ is replaced by $\mathcal{M}_0 = (\mathcal{M}_{x_1} \times \mathbb{R})$.
If in contrast $\partial_u \gamma_{\kappa_1}^1 = 0$ holds in a neighborhood of $\hat{\tilde{x}}_1$ on $\mathcal{M}_1$, only $\check{x}_1 := \tilde{x}_1$, $\check{u} := \tilde{u}$ and $\check{f}(\check{x}_1, \check{u}, x_2) := \tilde{f}(\tilde{x}_1, \tilde{u}, x_2)$ are replaced.

Step 2 κ_1 is increased by 1 and the new constraint $\gamma_{\kappa_1}^1 := L_{\check{f}}\gamma_{\kappa_1-1}^1$ is computed. All previous constraints are collected in the set $\gamma_1 := (\gamma_0^1, \ldots, \gamma_{\kappa_1-1}^1)$ and we assume that the rank of the Jacobian of γ_1 with respect to $\check{x}_1$ is constant in a neighborhood of $(\hat{\check{x}}_1, \hat{\check{u}})$ and

$$\operatorname{rk}(\partial_{\check{x}_1}\gamma_1) = \kappa_1 \tag{5}$$

holds. Then $\mathcal{M}_1 := \{\check{x}_1 \in \mathcal{M}_0 \mid \gamma_1 = 0\}$ is a submanifold.

Step 3 $\tilde{f} := \check{f}$, $\tilde{x}_1 := \check{x}_1$, and $\tilde{u} := \check{u}$ are replaced and the steps 1 and 2 are repeated until $\partial_{x_2}\gamma^1_{\kappa_1} \neq 0$ holds in a neighborhood of $\hat{\tilde{x}}_1$ on $\mathcal{M}_1$. This required κ_1 derivatives of the first algebraic equation.
The parameters $b^1_i \in \mathbb{R}$ are chosen such that the roots of the polynomial

$$P^1(\lambda) = (\lambda)^{\kappa_1} + \sum_{i=0}^{\kappa_1-1} b^1_i \cdot (\lambda)^i \tag{6}$$

are all in $\mathbb{C}^-$. With the implicit function theorem it is possible to find a fictitious feedback $\alpha_1 : (\mathcal{M}_1 \times \mathbb{R}^p \times \mathbb{R}^{m-1}) \to \mathbb{R}^m$ with $x_2 = \alpha_1(\tilde{x}_1, \tilde{u}, x_{2,1})$ and $x_{2,1} \in \mathbb{R}^{m-1}$ which solves

$$\gamma^1_{\kappa_1}(\tilde{x}_1, \tilde{u}, x_2) + \sum_{i=0}^{\kappa_1-1} b^1_i \gamma^1_i(\tilde{x}_1, \tilde{u}, x_2) = 0 \tag{7}$$

in a neighborhood of $\hat{\tilde{x}}_1$ on $\mathcal{M}_1$. The system Σ_0 controlled via $x_2 = \alpha_1(\tilde{x}_1, \tilde{u}, x_{2,1})$ is denoted by Σ_1. For this system $\mathcal{M}_1$ is attractive.
In the next steps the remaining algebraic equations beginning with the second are considered. Therefore, the variable $j = 2$ is introduced.

Step 4 Let $\dot{\tilde{x}}_1 = \tilde{f}(\tilde{x}_1, \tilde{u}, x_{2,(j-1)})$ with $x_{2,(j-1)} \in \mathbb{R}^{m-j+1}$ be the description of Σ_{j-1} and let $\mathcal{M}_j := \mathcal{M}_{j-1}$.

Step 5 If $\partial_{x_{2,(j-1)}}\gamma^j_{\kappa_j} \neq 0$ holds in a neighborhood of $\hat{\tilde{x}}_1$ on $\mathcal{M}_j$ we directly go to Step 7.
If $\partial_{\tilde{u}}\gamma^j_{\kappa_j} \neq 0$ holds in a neighborhood of $\hat{\tilde{x}}_1$ on $\mathcal{M}_j$, μ is increased by 1 and the system is dynamically extended by an integrator. Therefore, a regular input transformation $\tilde{u} = \beta_\mu(\tilde{x}_1, \tilde{v})$, $\tilde{v} \in \mathbb{R}^p$ is needed which locally satisfies $\operatorname{rank} \partial_{\tilde{v}}\beta_\mu = p$. The transformation is selected such that with the partitioning $\tilde{v} = (\tilde{v}_1, \tilde{v}_2)$, $\tilde{v}_1 \in \mathbb{R}^{p-1}$, $\tilde{v}_2 \in \mathbb{R}$ only $\tilde{v}_2$ has influence on $\gamma^j_{\kappa_j}$. Hence, locally $\partial_{\tilde{v}_1}(\gamma^j_{\kappa_j} \circ \beta_\mu) = 0$ and $\partial_{\tilde{v}_2}(\gamma^j_{\kappa_j} \circ \beta_\mu) \neq 0$ hold. In addition, β_μ is to be chosen such that at $\hat{\tilde{x}}_1$ $\partial_{x_1}\tilde{f} = (\partial_{\tilde{x}_1}\tilde{f} \circ \beta_\mu) \circ \beta_\mu^{-1}$, $\partial_{x_1}\gamma^j_{\kappa_j} = (\partial_{\tilde{x}_1}\gamma^j_{\kappa_j} \circ \beta_\mu) \circ \beta_\mu^{-1}$, and $\partial_{x_1}\gamma^i_0 = (\partial_{\tilde{x}_1}\gamma^i_0 \circ \beta_\mu) \circ \beta_\mu^{-1}$ is valid. Extending the state with $\tilde{v}_2$ leads to $\check{x}_1 := (\tilde{x}_1, \tilde{v}_2)$ and $\check{u} := (\check{u}_1, \check{u}_2)$ with $\check{u}_1 = \tilde{v}_1$. With $\dot{\tilde{v}}_2 = \check{u}_2$ it follows

$$\check{f}(\check{x}_1, \check{u}, x_{2,(j-1)}) := \begin{bmatrix} \tilde{f}(\tilde{x}_1, \beta_\mu(\tilde{x}_1, \tilde{v}), x_{2,(j-1)}) \\ \check{u}_2 \end{bmatrix}. \tag{8}$$

Moreover, $\mathcal{M}_{j-1}$ is replaced by $\mathcal{M}_{j-1} = \{(\tilde{x}_1, \tilde{v}_2) \in (\mathcal{M}_{j-2} \times \mathbb{R}) \mid \gamma_{j-1} = 0\}$.
If in contrast $\partial_{\tilde{u}}\gamma^j_{\kappa_j} = 0$ holds in a neighborhood of $\hat{\tilde{x}}_1$ on $\mathcal{M}_j$, only $\check{x}_1 := \tilde{x}_1$, $\check{u} := \tilde{u}$ and $\check{f}(\check{x}_1, \check{u}, x_{2,(j-1)}) := \tilde{f}(\tilde{x}_1, \tilde{u}, x_{2,(j-1)})$ are replaced.

Step 6 κ_j is increased by 1 and the new constraint $\gamma^j_{\kappa_j} := L_{\check{f}}\gamma^j_{\kappa_j-1}$ is computed. All previous constraints derived by the jth algebraic equation are collected in the set $\gamma_j := (\gamma^j_0, \ldots, \gamma^j_{\kappa_j-1})$ and we assume that the rank of the Jacobian of γ_j with respect to $\check{x}_1$ is constant in a neighborhood of $(\hat{\check{x}}_1, \hat{\check{u}})$ and

$$\operatorname{rk} \partial_{\check{x}_1} \begin{bmatrix} \gamma_1 \\ \vdots \\ \gamma_j \end{bmatrix} = \sum_{i=1}^{j} \kappa_i \, . \tag{9}$$

holds. Then $\mathcal{M}_j := \{\tilde{x}_1 \in \mathcal{M}_{j-1} \mid \gamma_j = 0\}$ is a submanifold.

Step 7 $\tilde{f} := \check{f}$, $\tilde{x}_1 := \check{x}_1$, and $\tilde{u} := \check{u}$ are replaced and the steps 5 and 6 are repeated until $\partial_{x_{2,(j-1)}}\gamma^j_{\kappa_j} \neq 0$ holds in a neighborhood of $\hat{\tilde{x}}_1$ on $\mathcal{M}_j$. This required κ_j derivatives of the jth algebraic equation.

The parameters $b^j_i \in \mathbb{R}$ are chosen such that the roots of the polynomial

$$P^j(\lambda) = (\lambda)^{\kappa_j} + \sum_{i=0}^{\kappa_j-1} b^j_i \cdot (\lambda)^i \tag{10}$$

are all in $\mathbb{C}^-$. With the implicit function theorem it is possible to find a fictitious feedback $\alpha_j : (\mathcal{M}_j \times \mathbb{R}^p \times \mathbb{R}^{m-j}) \to \mathbb{R}^{m-j+1}$ with $x_{2,(j-1)} = \alpha_j(\tilde{x}_1, \tilde{u}, x_{2,j})$ and $x_{2,j} \in \mathbb{R}^{m-j}$ which solves

$$\gamma^j_{\kappa_j}(\tilde{x}_1, \tilde{u}, x_{2,(j-1)}) + \sum_{i=0}^{\kappa_j-1} b^j_i \gamma^j_i(\tilde{x}_1, \tilde{u}, x_{2,(j-1)}) = 0 \tag{11}$$

in a neighborhood of $\hat{\tilde{x}}_1$ on $\mathcal{M}_j$. The system Σ_{j-1} controlled via $x_{2,(j-1)} = \alpha_j(\tilde{x}_1, \tilde{u}, x_{2,j})$ is denoted by Σ_j. For this system $\mathcal{M}_j$ is attractive.

Step 8 j is increased by 1 and steps 4–8 are repeated until $j = m$.

If the rank assumptions (5) and (9) in Step 2 and 6 are always satisfied, the algorithm terminates successfully. If the rank assumptions are not satisfied, the jth algebraic equation is not influenced by the remaining algebraic variables $x_{2,(j-1)}$. In this case, the algorithm will not terminate successfully. Later we will show that the algorithm will always terminate successfully for regular descriptor systems.

If the algorithm was successful, a new variable $x_e := \tilde{x}_1 \in \mathcal{M}_{x_e}$ is introduced where $\mathcal{M}_{x_e} = (\mathcal{M}_{x_1} \times \mathbb{R}^\mu)$ with $n_{x_e} := \dim \mathcal{M}_{x_e} = r + \mu$ because the system (3) was extended by μ integrators. The order of the dynamic extension is $n_u := \mu$ and the number of constraints $\kappa := \sum_{j=1}^{m} \kappa_j$.

All constraints are collected in

$$\gamma(x_e) = \begin{bmatrix} \gamma_0^1(\tilde{x}_1) \\ \vdots \\ \gamma_{\kappa_1-1}^1 \\ \gamma_0^2 \\ \vdots \\ \gamma_{\kappa_m-1}^m(\tilde{x}_1) \end{bmatrix}$$

and define the submanifold

$$\tilde{\mathcal{N}} = \{x_e \in \mathcal{M}_{x_e} \mid \gamma(x_e) = 0\} \ , \tag{12}$$

with the dimension $\dim \tilde{\mathcal{N}} = r + n_u - \kappa$. With the new input $v := \tilde{u}$ and the partitioning $x_e = (x_1, x_u)$, $x_u \in \mathbb{R}^{n_u}$, the dynamic extension

$$\dot{x}_u = r_x(x_u, x_1, v) \tag{13a}$$

$$u = r_u(x_u, x_1, v) \tag{13b}$$

is computed. The fictitious coupling controller

$$x_2 = \alpha_1\left(x_1, u, \alpha_2\left(\ldots \alpha_m(\tilde{x}_1, \tilde{u})\right)\right) = \alpha(x_e, v) \tag{14}$$

creates together with the dynamic extension the coupled state-space system

$$\dot{x}_e = \begin{bmatrix} f(x_1, \alpha(x_1, x_u, v), r_u(x_u, x_1, v)) \\ r_x(x_u, x_1, v) \end{bmatrix} = f_c(x_e, v) \tag{15a}$$

$$y = h(x_1, \alpha(x_1, x_u, v), r_u(x_u, x_1, v)) = h_c(x_e, v) \tag{15b}$$

which is equal to Σ_m. For this system $\tilde{\mathcal{N}}$ is invariant and attractive. Using (1b) the fictitious algebraic equation

$$0 = g(x_1, \alpha(x_u, x_1, v), r_u(x_u, x_1, v)) = g_c(x_e) \tag{16}$$

can be computed which is satisfied for all $x_e \in \tilde{\mathcal{N}}$ and $v \in \mathbb{R}^p$ and, therefore, does not depend on v.

Constraints that depend on x_2 are introduced using

$$\gamma_{\kappa_j}^j := L_{\check{f}} \gamma_{\kappa_j-1}^j \ , \quad \text{with } \check{f}(x_e, x_2, v) := \begin{bmatrix} f(x_1, x_2, r_u(x_u, x_1, v)) \\ r_x(x_u, x_1, v) \end{bmatrix}$$

which have to satisfy

$$\operatorname{rank} \partial_{x_2} \left[\gamma^j_{\kappa_j}\right]_{j \in \underline{m}} = m \, .$$

Equation (11) together with $z^j = \gamma^j_0, \dot{z}^j = \gamma^j_1, \ldots, (z^j)^{(\kappa_j)} = \gamma^j_{\kappa_j}, \, j \in \underline{m}$ describe a linear, autonomous differential equation

$$b^j_0 z^j + b^j_1 \dot{z}^j + \cdots + b^j_{\kappa_j - 1} (z^j)^{(\kappa_j - 1)} + (z^j)^{(\kappa_j)} = 0 \, .$$

Its dynamics is determined by the parameters b^j_i which are chosen in (10) such that $z^j = 0$ is a stable equilibrium point. As $z = 0$ holds on $\tilde{\mathcal{N}}$, the manifold $\tilde{\mathcal{N}}$ is invariant and attractive for (15). For proper descriptor systems this stabilization procedure is related to the simulation method of Baumgarte [5]. Many publications [3, 7] show the difficulty in choosing the stabilization parameters. However, simulating the coupled state-space system in an observer the numerical errors have minor influence compared to i.e. measurement errors.

Looking at all dynamic extensions for the descriptor system to get a proper extended descriptor system, the dynamic extension (13) is of minimal order as with every appearance of an input derivative only one additional variable is introduced.

In the following, we show that regularity of the descriptor system is necessary and sufficient for the successful termination of the algorithm and we derive conditions for the coupled state-space system (15) having the same solutions as the descriptor system (1).

Theorem 1 *Let* $(\hat{x}_1, \hat{x}_2)$ *be a consistent initial condition for the descriptor system* (1). *The algorithm terminates successfully in a neighborhood of* $(\hat{x}_e, \hat{v}) = (\hat{x}_1, \hat{x}_u, \hat{v})$ *with* $\hat{x}_2 = \alpha(\hat{x}_e, \hat{v})$ *and* $\hat{u} = r_u(\hat{x}_u, x_1, v)$ *if and only if the descriptor system* (1) *is regular in a neighborhood of* $(\hat{x}_1, \hat{x}_2, \hat{u})$.

Proof $\Leftarrow$: γ^j_i is the ith derivative of the jth algebraic equation of (1). Therefore, $\gamma^j_i = 0$ is fulfilled for all solutions of (1). As the descriptor system is regular, Lemma 1 leads to the invertibility of (3) with respect to $\langle z, x_2 \rangle$. Hence, all γ^j_i are independent (see [34]) and the rank assumptions (5) and (9) are satisfied. The algorithm terminates successfully.

$\Rightarrow$: Assume that the algorithm terminates successfully. Then a function (14) is computed which is unique on $\tilde{\mathcal{N}}$ independent of the choice of the parameters b^j_i. From $\hat{x}_e$, $\hat{v}$ and $z(t) = 0$ the function $x_2(t)$ can be determined uniquely with (15). Hence, (3) is invertible with respect to $\langle z, x_2 \rangle$ which is equivalent to the regularity of the descriptor system (1) because of Lemma 1. □

Theorem 2 *Let the descriptor system* (1) *be regular in a neighborhood of* $(\hat{x}_1, \hat{x}_2, \hat{u})$ *with* $(\hat{x}_1, \hat{x}_2)$ *being a consistent initial value.* $(x_1(t), x_2(t))$ *describes the solution of the descriptor system for this initial value and* $(\xi_1(t), \xi_u(t))$ *describes the solution of* (15) *for the initial value* $(\hat{x}_1, \hat{\xi}_u)$ *where* $\hat{\xi}_u$ *and* v *are chosen such that the output of* (13) *equals the input of the descriptor system* $u(t)$. *Then both systems* *have the same solutions and* $x_1(t) = \xi_1(t)$ *as well as* $x_2(t) = \alpha(\xi_1(t), \xi_u(t), v(t))$ *hold.*

Proof As (16) is satisfied for the chosen initial value, (16) is fulfilled for the solution of (15) for this initial value for all $v \in \mathbb{R}^p$. Therefore, (1b) for $\xi_1(t)$, $\xi_2(t)$ and $u(t)$ is always satisfied with $\xi_2(t) = \alpha(\xi_1(t), \xi_u(t), v(t))$. This is equivalent to $g(\xi_1(t), \xi_2(t), u(t)) = 0$. Because of the regularity of (1), $\xi_1(t)$ and $\xi_2(t)$ describe the solution of the descriptor system (1). □

At this point, we want to summarize the derived results and discuss further issues. If the descriptor system is proper, all constraints not depending on x_2 do not depend on the input u. Hence, no dynamic extension (13) is necessary. We want to emphasize that the system (15) is not a description in minimal coordinates for the descriptor system in general. This is only valid in the case of an index-1 descriptor system.

If the descriptor system is improper, a dynamic extension is computed during the algorithm. This dynamic extension has some nice properties.

Lemma 2 *For a given* $x_1(t)$ *the dynamic extension* (13) *is invertible and* $u(t)$ *determines* $x_u(t)$ *as well as* $v(t)$ *uniquely.*

Proof The proof is given by mathematical induction.

The first dynamic extension is given by

$$\begin{aligned} \dot{x}_u &= \check{u}_2 \\ u &= \beta_1(x_1, \tilde{v}) = r_u(x_u, x_1, \check{u}_1) \end{aligned}$$

with $x_u \in \mathbb{R}$ and $\tilde{v} = (\check{u}_1, x_u)$. Locally rank $\partial_{\tilde{v}} \beta_1 = \text{rank}\, \partial_{(\check{u}_1, x_u)} r_u = p$ holds, as the input transformation is regular. From the implicit function theorem it follows that the second equation is uniquely solvable for $(x_u, \check{u}_1) = \tilde{r}_u(x_1, u)$. $u(t)$ and $x_1(t)$ determine $\check{u}_1(t)$ and $x_u(t)$. Hence, $\check{u}_2 = \dot{x}_u$ fixes $\check{u}_2(t)$ and the dynamic extension of first order is invertible

Assume that the dynamic extension of kth order with $x_u \in \mathbb{R}^k$

$$\begin{aligned} \dot{x}_u &= r_x(x_u, x_1, \tilde{u}) \\ u &= r_u(x_u, x_1, \tilde{u}) \end{aligned}$$

is invertible for a given $x_1(t)$ and $u(t)$ determines $x_u(t)$ as well as $\tilde{u}(t)$ uniquely. The next dynamic extension leads to

$$\begin{aligned} \dot{x}_v &= \check{u}_2 \\ \tilde{u} &= \beta_k(x_1, x_u, \tilde{v}) = r_{\tilde{u}}(x_u, x_v, x_1, \check{u}_1) \end{aligned}$$

with $x_v \in \mathbb{R}$ and $\tilde{v} = (\check{u}_1, x_v)$. Locally rank $\partial_{\tilde{v}} \beta_k = \text{rank}\, \partial_{(\check{u}_1, x_v)} r_{\tilde{u}} = p$ holds. Again from the implicit function theorem it follows that the second equation is uniquely solvable for $(\breve{u}_1, x_v) = \tilde{r}_{\tilde{u}}(x_1, x_u, \tilde{u})$. Therefore, $\breve{u}_1(t)$ and $x_v(t)$ are fixed. $\breve{u}_2 = \dot{x}_v$ determines $\breve{u}_2(t)$.

It is shown by induction that the dynamic extension (13) is invertible for a given $x_1(t)$ and $u(t)$ is uniquely determined by $x_u(t)$ and $v(t)$. □

4 Causal Observability and Observer Design

A dynamic system is called an observer for the descriptor system (1) if it satisfies both following properties:

(i) (Consistence) if $\hat{x}_b = \hat{x}$, $x_b(t) = x(t)$ holds for all $t > t_0$ and
(ii) (Convergence) if $\hat{x}_b \neq \hat{x}$, $\lim\limits_{t\to\infty} x_b(t) - x(t) = 0$ is fulfilled.

The observer only takes the input u and output y of the descriptor system. In many cases, the observers for descriptor systems are designed as descriptor systems themselves. Avoiding problems concerning the simulation of the observer, it is reasonable to design the observer as state-space system

$$\dot{\xi}_b = a_b(\xi_b, u, y) \tag{17a}$$

$$x_b = c_b(\xi_b, u, y) \tag{17b}$$

where $\xi_b \in \mathbb{R}^{n_b}$ is the state of the observer whose dimension n_b is not known in advance. A dynamic system is a *causal observer* if in addition to the properties of an observer the following property is satisfied:

(iii) (Causality) the observer can be described by a state-space system (17).

If such an observer exists, the descriptor system (1) is called *causal observable*. Before going into details of causal observability, local observability is introduced.

The following definition of local observability is taken from [32]. As the definition is independent of the system description, it can be used for state-space systems as well as for descriptor systems. $y(t, t_0, \xi_0, u)$ describes the output function of (1) for the solution of (1) with the consistent initial condition $\xi_0 := x(t_0)$ and the input u. In addition, the regularity of the descriptor system is assumed.

Definition 4 ([32, Definition 3.28]) Let $V \subset \mathscr{M}_x$ be an open set containing ξ_1 as well as ξ_2. ξ_1 and ξ_2 are called *V-indistinguishable*, denoted by $\xi_1 \mathrm{I}^V \xi_2$, if for every admissible constant control $u \in U_u$ the solutions $\xi(t, t_0, \xi_1, u)$ and $\xi(t, t_0, \xi_2, u)$ for $t \leq t_1$, with $t_1 > t_0$ arbitrary, remain in V and the output functions $y(t, t_0, \xi_1, u)$, $t \geq t_0$, and $y(t, t_0, \xi_2, u)$, $t \geq t_0$ are the same for $t \leq t_1$ on their common domain of definition. The descriptor system (1) is called *locally observable at* $\hat{\xi}$ if there exists a neighborhood W of $\hat{\xi}$ such that for every neighborhood $V \subset W$ of $\hat{\xi}$ the relation $\xi_1 \mathrm{I}^V \xi_2$ implies that $\xi_1 = \xi_2$. If the system is locally observable at each $\hat{\xi}$, then it is called *locally observable*.

For linear descriptor systems different observability concepts are distinguished (see e.g. [9]). The presented definition of local observability equals the observability of the slow subsystem for linear descriptor systems. observability concepts for the impulse behavior are not relevant for this contribution.

For a proper descriptor system the coupled state-space system (15) can directly be used as process copy in state-space form because $v = u$. If (15) is locally observable, a state-space observer can be designed. As the solutions of the coupled state-space system and the descriptor system are the same on $\tilde{\mathcal{N}}$ according to Theorem 2 the observer is also a causal observer for the descriptor system. The properness of the descriptor system and the local observability of (15) are obviously sufficient, but they are not necessary at all.

Transferring the descriptor observer design to a state-space observer design constitutes the basic idea for further considerations and will finally lead to necessary and sufficient conditions. Therefore, we seek a causal process copy of the descriptor system for which a state-space observer can be computed. In the case of an improper descriptor system, (15) is a causal process copy, but the input v only can be calculated from u by differentiation. As it is not possible to describe an ideal differentiator in state-space form this will not lead to a causal observer.

However, if it is possible to determine v only by y and u, then the input of (15) is known and (15) can be used as causal process copy. But u is not visible in (15) anymore. In the algorithm the dynamic extension (13) was generated. The output of (13) can be interpreted as additional output equation as u will be known by the observer. Furthermore, all solutions of the descriptor system lie on $\tilde{\mathcal{N}}$ so that $z = 0$ is also known. In summary, we have the coupled state-space system with extended output equation

$$\dot{x}_{\mathrm{e}} = f_{\mathrm{c}}(x_{\mathrm{e}}, v) \tag{18a}$$

$$y_{\mathrm{e}} = \begin{bmatrix} y \\ u \\ z \end{bmatrix} = \begin{bmatrix} h_{\mathrm{c}}(x_{\mathrm{e}}, v) \\ r_{\mathrm{u}}(x_{\mathrm{e}}, v) \\ g_{\mathrm{c}}(x_{\mathrm{e}}) \end{bmatrix} = h_{\mathrm{e}}(x_{\mathrm{e}}, v)\,. \tag{18b}$$

It is possible to derive an observability result for (18) from the local observability of the descriptor system.

Lemma 3 *Let the descriptor system* (1) *be locally observable at* $(\hat{x}_1, \hat{x}_2)$ *for inputs in a neighborhood of* $\hat{u}$ *and let* $\hat{u} = r_{\mathrm{u}}(\hat{x}_{\mathrm{u}}, \hat{x}_1, \hat{v})$ *hold. Two solutions* $x_{\mathrm{e},1}(t)$ *and* $x_{\mathrm{e},2}(t)$ *of the system* (18) *continuing in the neighborhood of* $(\hat{x}_1, \hat{x}_{\mathrm{u}})$ *agree if* $y_{\mathrm{e},1}(t) = y_{\mathrm{e},2}(t)$, $u_1(t) = \mathrm{const}$ *as well as* $u_2(t) = \mathrm{const}$ *hold.*

Proof Assume there exist two solutions $x_{\mathrm{e},1}(t) \neq x_{\mathrm{e},2}(t)$ of (18) for the inputs $v_1(t)$ and $v_2(t)$, respectively, such that $y_{\mathrm{e},1}(t) = y_{\mathrm{e},2}(t)$, $u_1(t) = \mathrm{const}$ as well as $u_2(t) = \mathrm{const}$ hold. $y_{\mathrm{e},1}(t) = y_{\mathrm{e},2}(t)$ results in $u_1(t) = u_2(t)$ as u is included in y_{e}. Because of the local observability of the descriptor system, with the partitioning $x_{\mathrm{e,i}} = (x_{1,i}, x_{\mathrm{u},i})$ it follows $x_{1,1} = x_{1,2}$. Because of Lemma 2, $u(t)$ together with $x_1(t)$ determine uniquely $x_{\mathrm{u}}(t)$ and $v(t)$, $u_1(t) = u_2(t)$ leads to $x_{\mathrm{u},1}(t) = x_{\mathrm{u},2}(t)$ which contradicts the assumption. □

Assuming that

$$\operatorname{rk} \partial_v h_e(x_e, v) = p \tag{19}$$

is satisfied in a neighborhood of $(\hat{x}_e, \hat{v})$ the extended output equation of (18) can be uniquely solved for v. This leads to

$$v = r_b(x_e, y_e) \,. \tag{20}$$

Hence, the input of (18) is known and the observer can be designed using

$$\dot{x}_e = f_c(x_e, r_b(x_e, y_e)) \tag{21a}$$

$$\tilde{y}_e = h_e(x_e, r_b(x_e, y_e)) \tag{21b}$$

if (18) is locally observable. Therefore, the differential equation (17a) of the observer is designed. Computing x_b from ξ_b the output equation (17b) of the observer has to be found. The first r components of ξ_b, i.e. $\xi_{b,1}$, directly provide the estimate for x_1. Using (14) and (20) yield

$$x_b = \begin{bmatrix} \xi_{b,1} \\ \tilde{\alpha}(\xi_b, r_b(\xi_b, y_e)) \end{bmatrix} . \tag{22}$$

In Fig. 1 the derived systems and particular transformations are illustrated to summarize the previous results. In order for (17) to be a causal observer, the three

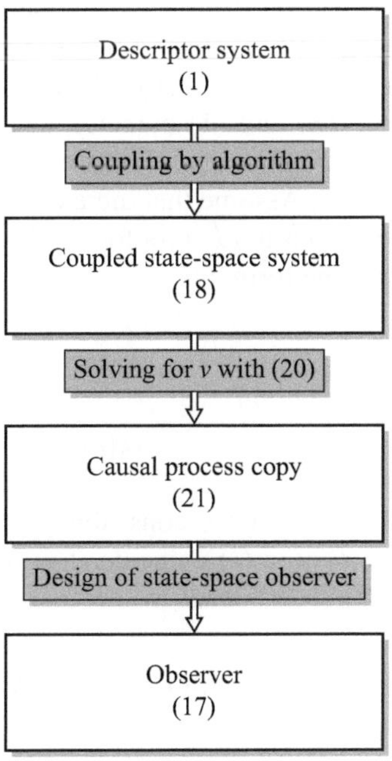

Fig. 1 Summary of the observer design

properties introduced at the beginning of the section have to be fulfilled. As (17) is a state-space system, the *causality* is trivially satisfied.

As (20) solves the extended output equation exactly for the input v, $f_c(x_e, v) = f_c(x_e, r_b(x_e, h_e(x_e, v)))\ \forall x_e \in U \subset \tilde{\mathcal{N}}, v \in V \subset \mathbb{R}^p$ holds. The solutions of (15) for consistent initial conditions always remain in $\tilde{\mathcal{N}}$. Together with Theorem 2 the *consistence* property is satisfied.

For the *convergence* we show in the following that (21) is locally observable. Then the design of a state-space observer is possible. Converging to the solution of (21) is equivalent to converging to the solution of (1) with Theorem 2.

Lemma 4 *Let the descriptor system* (1) *be locally observable at* $(\hat{x}_1, \hat{x}_2)$ *for inputs in a neighborhood of* $\hat{u}$ *and let* $\hat{u} = r_u(\hat{x}_u, \hat{x}_1, \hat{v})$ *hold. If the condition* (19) *is satisfied in a neighborhood of* $(\hat{x}_1, \hat{x}_u, \hat{v})$, *the system* (21) *is locally observable at* $(\hat{x}_1, \hat{x}_u)$.

Proof Assume that the system (21) is not locally observable. There exist two solutions $x_{e,1}(t) \neq x_{e,2}(t)$ for the constant input $y_{e,1}(t) = y_{e,2}(t)$ with the outputs satisfying $\tilde{y}_{e,1}(t) = \tilde{y}_{e,2}(t)$. With (20) unique $v_1(t)$ and $v_2(t)$ can be found. Applying Lemma 3 leads to $x_{e,1}(t) = x_{e,2}(t)$ which contradicts the assumption. Hence, the system (18) is locally observable. □

In addition it is possible to show that the condition (19) is not only sufficient for causal observability but it is also necessary for locally observable descriptor systems.

Theorem 3 *Let the regular descriptor system* (1) *be locally observable at* $\zeta_0 \in V \subset \tilde{\mathcal{N}}$. *The descriptor system* (1) *is causal observable if and only if the condition* (19) *is satisfied in* V.

Proof $\Leftarrow$: If the regular descriptor system (1) is locally observable and the condition (19) is satisfied because of Lemma 4, the system (21) is a locally observable causal process copy. Hence, a causal observer can be designed. As the solutions of (18) and (1) coincide using Theorem 2 and as (20) computes v exactly, the observer is a causal observer for (1).

$\Rightarrow$: Assume that there exists a dynamic system (17) which fulfills the properties of a causal observer for the descriptor system (1). The solution $x_b(t)$ of the observer has the form

$$\xi_b(t) = \xi_b(t_0) + \int_{t_0}^{t} a_b(\xi_b(\tau), u(\tau), y(\tau)) \mathrm{d}\tau$$

$$x_b(t) = c_b\left(\xi_b(t), u(t), y(t)\right) .$$

Because of the consistency of the observer with $\hat{x}_b = x_b(t_0) = \hat{x}$ the solutions $x_b(t) = x(t)$ coincide. The partitioning $c_b = (c_{b,1}, c_{b,2})$ with $c_{b,2} : (\mathcal{M}_\xi \times \mathbb{R}^p \times \mathbb{R}^q) \rightarrow \mathbb{R}^m$ leads to

$$c_{b,2}(\xi_b, u, y) = c_{b,2}(\xi_b, h_e(x_e, v)) \stackrel{!}{=} x_2 = \alpha(x_e, v) \quad \forall x_e \in \tilde{\mathcal{N}}, v \in \mathbb{R}^p .$$

As $v \in \mathbb{R}^p$ can be chosen arbitrarily, this equation can only be satisfied if

$$\frac{\partial(c_{b,2} \circ h_e)}{\partial v} = \frac{\partial \alpha}{\partial v}$$

holds. Assume that the condition (19) is violated. Then there exists a v^i fulfilling

$$\frac{\partial h_e(x_e, v)}{\partial v^i} = 0$$

possibly requiring a regular input transformation. Equation (18b) leads to $\partial_{v^i} r_u = 0$. From Lemma 2 (13) is invertible resulting in $\partial_{v^i} r_x \neq 0$. As the extension by an integrator in the algorithm is only carried out if the input appears in a restriction, there is at least one $\gamma^j_{\kappa_j}$ depending on v^i. Hence, it follows that $\partial_{v^i} \alpha \neq 0$ and, therefore,

$$\frac{\partial(c_{b2} \circ h_e)}{\partial v^i} \neq \frac{\partial \alpha}{\partial v^i}$$

holds which violates the consistency property of the observer. □

The last theorem provides a new condition for causal observability that is not only sufficient but also necessary. To check the condition only the coupled state-space system (15) is required which is also used in the observer design procedure. This design procedure for causal observers for all causal observable nonlinear descriptor systems is thereby transferred to the design of a state-space observer for a locally observable state-space system. For the actual state-space observer design arbitrary methods from the literature can be used, what is not discussed further. If the observer state converges to the state of (21), it converges to the state of (15). With Theorem 2 it also converges to the descriptor of the descriptor system (1) using (22).

For proper descriptor systems the observer design and the condition of causal observability simplifics.

Theorem 4 *A regular, proper descriptor system* (1) *is causal observable if it is locally observable at* $\hat{x}$.

Proof If the descriptor system is proper, the matrix $\partial_v r_u$ has full rank and (19) is satisfied independent of h_c. The causal observability is a direct consequence of Theorem 3. □

For linear systems necessary and sufficient conditions for causal observability are known (see e.g. [1]). In [25] the authors show that the new condition equals the familiar conditions if the observability of the descriptor system is replaced by the detectability of the slow subsystem.

5 Examples

To exemplify the presented approach we design a causal observer for two models. First model is the descriptor system of a linear three-mass oscillator and the second one is an academic nonlinear example.

5.1 Three-Mass Oscillator

Figure 2 shows a system consisting of three masses coupled via two stiff rods. The length of the rod between m_1 and m_2 is of variable length. m_1 and m_3 are connected to the environment via spring and damper elements. The positions of the masses are denoted by x^1, x^3, x^5 and their velocities by x^2, x^4, x^6, respectively. Together they form the dynamic variable x_1. The algebraic variable x_2 corresponds to the force between the first two masses x^7 and the force x^8 between m_2 and m_3. The first input, the force u^1, acts on the third mass. The second input u^2 equals the distance between masses m_1 and m_2. The system is described by the descriptor system

$$\dot{x}_1 = \begin{bmatrix} x^2 \\ -\frac{c_1}{m_1}x^1 - \frac{d_1}{m_1}x^2 + \frac{1}{m_1}x^7 \\ x^4 \\ -\frac{1}{m_2}x^7 + \frac{1}{m_2}x^8 \\ x^6 \\ -\frac{c_2}{m_3}x^5 - \frac{d_2}{m_3}x^6 - \frac{1}{m_3}x^8 + \frac{1}{m_3}u^1 \end{bmatrix} \qquad 0 = \begin{bmatrix} -x^1 + x^3 + u^2 \\ -x^3 + x^5 \end{bmatrix}$$

$$y = x^7 .$$

The second input makes the system improper and the slow subsystem of the descriptor system is observable. In contrast to the two-mass oscillator in [25] this system is not impulse observable. The parameters are collected in Table 1.

It seems strange having a real system that is improper. In this case, the reason is the simplification in the modeling. It is not possible to change positions abruptly in reality although u^2 is an input. In a real system u^2 would be the output of an actuator.

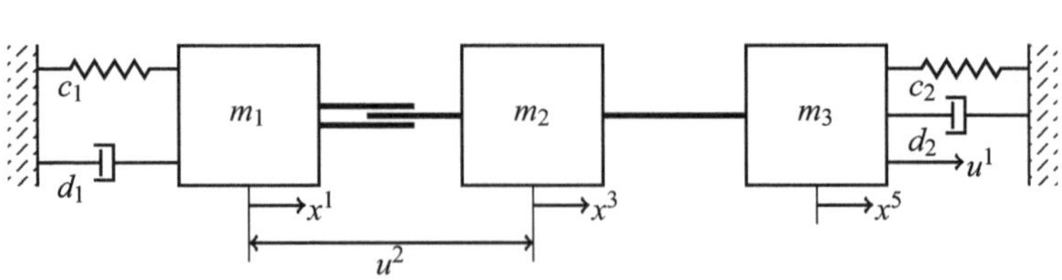

Fig. 2 Three-mass oscillator

Table 1 Parameters of the three-mass oscillator

Symbol	Value	Unit
c_1	1	N/m
c_2	2	N/m
d_1	0.1	Ns/m
d_2	0.3	Ns/m
m_1	1	kg
m_2	0.1	kg
m_3	0.2	kg

If the function $u^2(t)$ is sufficiently smooth and can be produced by the actuator, the descriptor system shows the correct behavior.

With the parameters of Table 1 and $b_0^i = 25$, $b_1^i = 10$, $b_2^i = 1$ for $i = 1, 2$ the algorithm creates the coupled state-space system

$$\dot{x}_e = \begin{bmatrix} 0 & 1 & 0 & 0 & 0 & 0 & 0 & 0 \\ -6.54 & -2.38 & 1.92 & 0.77 & 2.31 & 1.31 & 5.77 & 2.31 \\ 0 & 0 & 0 & 1 & 0 & 0 & 0 & 0 \\ 18.46 & 7.62 & -23.08 & -9.23 & 2.31 & 1.31 & -19.23 & -7.69 \\ 0 & 0 & 0 & 0 & 0 & 1 & 0 & 0 \\ 18.46 & 7.62 & 1.92 & 0.77 & -22.69 & -8.69 & -19.23 & -7.69 \\ 0 & 0 & 0 & 0 & 0 & 0 & 0 & 1 \\ 0 & 0 & 0 & 0 & 0 & 0 & 0 & 0 \end{bmatrix} x_e$$

$$+ \begin{bmatrix} 0 & 0 \\ 0.77 & 0.23 \\ 0 & 0 \\ 0.77 & -0.77 \\ 0 & 0 \\ 0.77 & -0.77 \\ 0 & 0 \\ 0 & 1 \end{bmatrix} v \tag{23a}$$

with extended output equation

$$y_e = \begin{bmatrix} -5.54 & -2.28 & 1.92 & 0.77 & 2.31 & 1.31 & 5.77 & 2.31 \\ 0 & 0 & 0 & 0 & 0 & 0 & 0 & 0 \\ 0 & 0 & 0 & 0 & 0 & 0 & 1 & 0 \\ -1 & 0 & 1 & 0 & 0 & 0 & 1 & 0 \\ 0 & 0 & -1 & 0 & 1 & 0 & 0 & 0 \end{bmatrix} x_e + \begin{bmatrix} 0.77 & 0.23 \\ 1 & 0 \\ 0 & 0 \\ 0 & 0 \\ 0 & 0 \end{bmatrix} v\,. \tag{23b}$$

As the condition (19) is satisfied, the descriptor system is causal observable. Solving the output equation for v leads to

$$v = \begin{bmatrix} 0 & 0 & 0 & 0 & 0 & 0 & 0 & 0 \\ 24 & 9.9 & -8.33 & -3.33 & -10 & -5.67 & -25 & -10 \end{bmatrix} x_e + \begin{bmatrix} 0 & 1 & 0 & 0 & 0 \\ 4.33 & -3.33 & 0 & 0 & 0 \end{bmatrix} y_e \,.$$

Inserting this function into the coupled state-space system creates the causal process copy

$$\dot{x}_e = \begin{bmatrix} 0 & 1 & 0 & 0 & 0 & 0 & 0 & 0 \\ -1 & -0.1 & 0 & 0 & 0 & 0 & 0 & 0 \\ 0 & 0 & 0 & 1 & 0 & 0 & 0 & 0 \\ 0 & 0 & -16.67 & -6.67 & 10 & 5.67 & 0 & 0 \\ 0 & 0 & 0 & 0 & 0 & 1 & 0 & 0 \\ 0 & 0 & 8.33 & 3.33 & -15 & -4.33 & 0 & 0 \\ 0 & 0 & 0 & 0 & 0 & 0 & 0 & 1 \\ 24 & 9.9 & -8.33 & -3.33 & -10 & -5.67 & -25 & -10 \end{bmatrix} x_e$$

$$+ \begin{bmatrix} 0 & 0 & 0 & 0 & 0 \\ 1 & 0 & 0 & 0 & 0 \\ 0 & 0 & 0 & 0 & 0 \\ -3.33 & 3.33 & 0 & 0 & 0 \\ 0 & 0 & 0 & 0 & 0 \\ -3.33 & 3.33 & 0 & 0 & 0 \\ 0 & 0 & 0 & 0 & 0 \\ 4.33 & -3.33 & 0 & 0 & 0 \end{bmatrix} y_e \qquad (24a)$$

$$\tilde{y}_e = \begin{bmatrix} 0 & 0 & 0 & 0 & 0 & 0 & 0 & 0 \\ 0 & 0 & 0 & 0 & 0 & 0 & 0 & 0 \\ 0 & 0 & 0 & 0 & 0 & 0 & 1 & 0 \\ -1 & 0 & 1 & 0 & 0 & 0 & 1 & 0 \\ 0 & 0 & -1 & 0 & 1 & 0 & 0 & 0 \end{bmatrix} x_e + \begin{bmatrix} 1 & 0 & 0 & 0 & 0 \\ 0 & 1 & 0 & 0 & 0 \\ 0 & 0 & 0 & 0 & 0 \\ 0 & 0 & 0 & 0 & 0 \\ 0 & 0 & 0 & 0 & 0 \end{bmatrix} y_e \,. \qquad (24b)$$

It is possible to design a Luenberger observer for this system whose eigenvalues are at $(-0.5, -1, -1.5, -2, -2.5, -3, -3.5, -4)$. The simulation of the descriptor system is shown in Fig. 3a with the initial value

$$x(0) = \begin{bmatrix} -0.17 & 2.06 & -1.47 & 2.06 & -1.47 & 2.06 & 1.52 & 1.79 \end{bmatrix}^{\mathrm{T}}$$

and the input function

$$u(t) = \begin{bmatrix} 0.1t \\ 1 + 0.3\cos(2t) \end{bmatrix}.$$

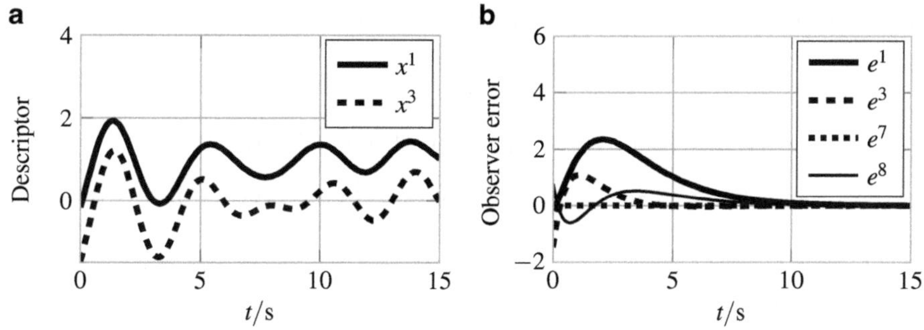

Fig. 3 Three-mass oscillator and observer error. (**a**) Simulation of the three-mass oscillator. (**b**) Simulation of the observer error

Figure 3b shows the observer error for the initial value $\xi_b(0) = 0$ of the observer. The asymptotic decay of the observer error is clearly illustrated.

The three-mass oscillator exemplifies the design of causal observers for a linear system. Although, the design was comparatively simple, the basic idea is visualized. With the next example a more challenging observer design is shown.

5.2 *Academic Nonlinear Example*

We consider the descriptor system

$$\dot{x}_1 = \begin{bmatrix} u^1 + \cos(x^1) \\ u^2 + x^3 \end{bmatrix} \qquad 0 = x^2 + x^1 u^1 + u^2 \qquad y = x^3 \tag{25}$$

with the differential variable $x_1 = (x^1, x^2)$, the algebraic variable $x_2 = x^3$, the input $u = (u^1, u^2)$, and the output y. For inputs satisfying $u(t_0) = (\hat{u}^1, \hat{u}^2)$

$$\hat{x} = (\arccos(-\hat{u}^1), -\arccos(-\hat{u}^1)\hat{u}^1 - \hat{u}^2, -\hat{u}^2)$$

is a consistent initial value and an equilibrium point. In the following, we take $\hat{u}^1 = 0.5$ and $\hat{u}^2 = 0$ and consider the descriptor system locally in a neighborhood of $\hat{x}$. A short analysis shows that $\hat{x}$ is an asymptotic stable equilibrium point and that the system is locally observable at $\hat{x}$. Using the algorithm with $b_0^1 = 1$ leads to

$$\gamma_0^1 = x^2 + x^1 u^1 + u^2$$

An input transformation is obviously necessary because γ_0^1 depends on u^1 as well as u^2. The regular transformation

$$\beta(x_1, \tilde{v}) = \begin{bmatrix} \tilde{v}^1 \\ x^1 \tilde{v}^2 - x^1 v^1 \end{bmatrix}$$

satisfies the requirements and together with $x_u = \tilde{v}_2$, $v^1 = \tilde{v}^1$ creates

$$\begin{aligned}
\gamma_0^1 &= x^2 + x^1 x_u \\
\gamma_1^1 &= x^3 - x^1\tilde{v}^1 + x^1 x_u + x^1\tilde{v}^2 + x_u(v^1 + \cos(x^1)) \\
\alpha &= x^1 v^1 - x^2 - x^1 v^2 - x_u v^1 - 2x^1 x_u - x_u \cos(x^1) \\
\dot{x}_u &= v_2 \,.
\end{aligned}$$

This leads to the coupled state-space system

$$\begin{aligned}
\dot{x}_e &= \begin{bmatrix} \cos(x^1) + v^1 \\ -x^2 - x^1 v^2 - x_u x_1 - x_u v^1 - x_u \cos(x^1) \\ v^2 \end{bmatrix} \\
y_e &= \begin{bmatrix} x^1 v^1 - x^2 - x^1 v^2 - x_u v^1 - 2x^1 x_u - x_u \cos(x^1) \\ v^1 \\ x^1 x_u - x^1 v^1 \\ x^2 + x^1 x_u \end{bmatrix}
\end{aligned}$$

whose extended output equation can be solved for v with

$$v = \begin{bmatrix} u^1 \\ -\frac{1}{x^1}\left(x^2 + y - x^1 u^1 + 2x^1 x_u + x_u u^1 + x_u \cos(x^1)\right) \end{bmatrix}.$$

Inserting this function into the coupled state-space system results in the causal process copy

$$\begin{aligned}
\dot{x}_e &= \begin{bmatrix} \cos(x^1) + u^1 \\ y - x^1 u^1 + x^1 x_u \\ u^1 - 2x_u - \frac{1}{x^1}\left(x^2 + y + x_u u^1 + x_u \cos(x^1)\right) \end{bmatrix} = f_k(x_e, y_e) \\
\tilde{y}_e &= \begin{bmatrix} y \\ u^1 \\ x^1 x_u - x^1 u^1 \\ x^2 + x^1 x_u \end{bmatrix} = h_k(x_e, y_e) \\
x_b &= \begin{bmatrix} x^1 \\ x^2 \\ y \end{bmatrix} = c_b(x_e, y_e)
\end{aligned}$$

which is locally observable. The linear output injection with the observer matrix

$$L = \begin{bmatrix} 0 & 0 & 8 & -6 \\ 0 & 0 & -8 & 8 \\ 0 & 0 & 2 & -1 \end{bmatrix}$$

provides an observer

$$\dot{\xi}_{\mathrm{b}} = f_{\mathrm{k}}(\xi_{\mathrm{b}}, y_{\mathrm{e}}) + L(y_{\mathrm{e}} - h_{\mathrm{k}}(\xi_{\mathrm{b}}, y_{\mathrm{e}})) \tag{26a}$$

$$x_{\mathrm{b}} = c_{\mathrm{b}}(\xi_{\mathrm{b}}, y_{\mathrm{e}}) \tag{26b}$$

for the causal process copy which can be used as observer for the descriptor system. The observer has an asymptotically stable observer-error dynamics at $\hat{\xi}_{\mathrm{b}} = (\hat{x}, \hat{u}^1 + \hat{u}^2/\hat{x}^1)$.

The descriptor system is simulated with the consistent initial value $\hat{x}$ at time $t_0 = 0$. The input function

$$u(t) = \begin{bmatrix} \hat{u}^1 + 0.1\sin(t) \\ \hat{u}^2 + 0.01\sin(2t) \end{bmatrix}$$

is chosen. The observer is initialized with $\xi_{\mathrm{b}}^1(t_0) = \xi_{\mathrm{b}}^2(t_0) = \xi_{\mathrm{b}}^3(t_0) = 0.1$.

Figure 4a shows the simulation of the descriptor system and the decay of the observer error is depicted in Fig. 4b. Although the observer is valid only locally, the observer converges after a considerably large initial error.

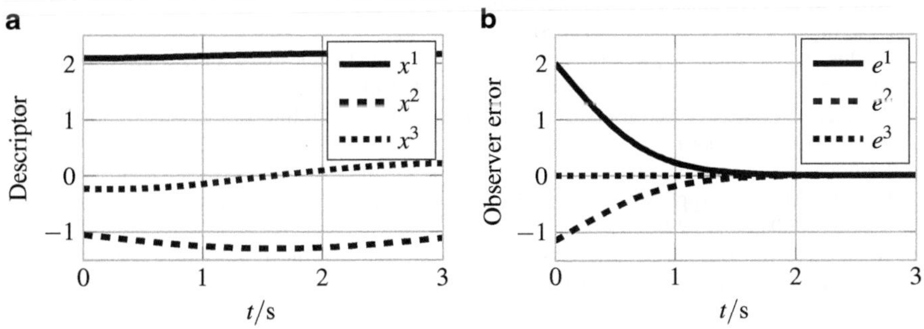

Fig. 4 Simulation of the descriptor system (25) and error. (**a**) Simulation of the descriptor system (25). (**b**) Error $e^i = x^i - x_{\mathrm{b}}^i$ of the observer (26)

Conclusion

This contribution is concerned with the design of causal observers for descriptor systems. At first, an algorithm was presented to transform the descriptor system into a coupled state-space system. On a submanifold of the state manifold both have the same solutions. Therefore, the coupled state-space system is adequate for the design procedure. As the input of the descriptor system and the state-space system are different in general, the coupled state-space system has to be modified such that the modified system has known inputs. If this is possible, the corresponding condition turns out to be a new necessary and sufficient condition for causal observability of locally observable descriptor systems. The actual observer design is carried out with the modified system to create a state-space observer. It is shown that this observer is a causal observer for the original descriptor system.

At the end, two examples illustrate the new approach. The first example is a three-mass oscillator which gives a comprehensive insight in the method as it is a linear system. The second example is an academic nonlinear system. It exemplifies some issues arising in nonlinear systems. Both examples show the capability of the new method.

References

1. Ailon, A.: On the reduced-order causal observer for generalized control systems. Int. J. Control **57**(6), 1311–1323 (1993)
2. Aliyu, M.D.S., Boukas, E.K.: Kalman filtering for affine nonlinear descriptor systems. Circuits Syst. Signal Process. **30**, 125–142 (2011)
3. Ascher, U.M., Chin, H., Reich, S.: Stabilization of DAEs and invariant manifolds. Numer. Math. **67**, 131–149 (1994). http://dx.doi.org/10.1007/s002110050020
4. Åslund, J., Frisk, E.: An observer for non-linear differential-algebraic systems. Automatica **42**(6), 959–965 (2006)
5. Baumgarte, J.: Stabilization of constraints and integrals of motion in dynamical systems. Comput. Methods Appl. Mech. Eng. **1**(1), 1–16 (1972). http://www.sciencedirect.com/science/article/pii/0045782572900187
6. Boutayeb, M., Darouach, M.: Observers design for nonlinear descriptor systems. In: Proceedings of the 34th IEEE Conference on Decision and Control, New Orleans, vol. 3, pp. 2369–2374 (1995)
7. Burgermeister, B., Arnold, M., Esterl, B.: DAE time integration for real-time applications in multi-body dynamics. ZAMM – Z. Angew. Math. Mech. **86**(10), 759–771 (2006). http://dx.doi.org/10.1002/zamm.200610284
8. Campbell, S.L., Delebecque, F., Nikoukhah, R.: Observer design for linear time varying descriptor systems. In: Proceedings Control of Industrial Systems, Belfort, vol. 1, pp. 507–512 (1997)
9. Dai, L.: Singular Control Systems. Springer, Berlin/Heidelberg (1989)
10. Darouach, M., Boutat-Baddas, L.: Observers for a class of nonlinear singular systems. IEEE Trans. Autom. Control **53**(11), 2627–2633 (2008)

11. Darouach, M., Boutat-Baddas, L.: Observers for Lipschitz nonlinear descriptor systems: application to unknown inputs systems. In: Proceedings of the 16th IEEE Mediterranean Conference on Control and Automation, Ajaccio, pp. 1369–1374 (2008)
12. Darouach, M., Boutayeb, M.: Design of observers for descriptor systems. IEEE Trans. Autom. Control **40**(7), 1323–1327 (1995)
13. Fahmy, M.M., O'Reilly, J.: Observers for descriptor systems. Int. J. Control **49**(6), 2013–2028 (1989)
14. Frisk, E., Åslund, J.: An observer for semi-explicit differential-algebraic systems. In: Proceedings of the 16th IFAC World Congress (IFAC), Prague, pp. 671–676. Elsevier (2005)
15. Gao, Z., Ho, D.W.: State/noise estimator for descriptor systems with application to sensor fault diagnosis. IEEE Trans. Signal Process. **54**(4), 1316–1326 (2006)
16. Gear, C.W.: Differential-algebraic equation index transformations. SIAM J. Numer. Anal. **9**, 39–47 (1988)
17. Gerdin, M.: Local identifiability and observability of nonlinear differential-algebraic equations. In: Proceedings of the 14th IFAC Symposium on System Identification, Newcastle, vol. 14, pp. 802–807 (2006)
18. Hou, M., Müller, P.C.: Design of a class of Luenberger observers for descriptor systems. IEEE Trans. Autom. Control **40**, 133–136 (1995)
19. Hou, M., Müller, P.C.: Causal observability of descriptor systems. IEEE Trans. Autom. Control **44**, 158–163 (1999)
20. Hou, M., Müller, P.C.: Observer design for descriptor systems. IEEE Trans. Autom. Control **44**, 164–168 (1999)
21. Hou, M., Müller, P.C.: Observer design for a class of descriptor systems. In: Proceedings of the European Control Conference, Cambridge, 1–4 Sept 2003. CD–Rom
22. Kaprielian, S., Turi, J.: An observer for a nonlinear descriptor system. In: Proceedings of the 31st IEEE Conference on Decision and Control, Tucson, vol. 1, pp. 975–976 (1992)
23. Koenig, D.: Observer design for unknown input nonlinear descriptor systems via convex optimization. IEEE Trans. Autom. Control **51**(6), 1047–1052 (2006)
24. Labisch, D.: Verkopplungsbasierte Methoden zum Regler- und Beobachterentwurf für nichtlineare Deskriptorsysteme. Fortschritt-Berichte VDI, Reihe 8, Nr. 1227, VDI-Verlag, Düsseldorf (2013)
25. Labisch, D., Konigorski, U.: Entwurf kausaler Beobachter für lineare Deskriptorsysteme. In: Adamy, J., Sawodny, O. (eds.) GMA Fachausschuss 1.40 "Theoretische Verfahren der Regelungstechnik", Workshop, 16–19 Sept 2011, Salzburg, pp. 289–301 (2012)
26. Labisch, D., Manderla, M., Konigorski, U.: Control, simulation and stability analysis of nonlinear regular proper DAEs. In: Proceedings of the 20th Mediterranean Conference on Control and Automation, Barcelona, pp. 392–397 (2012)
27. Lu, G., Ho, D.W.: Full-order and reduced-order observers for Lipschitz descriptor systems: the unified LMI approach. IEEE Trans. Circuits Syst. II: Express Briefs **53**(7), 563–567 (2006)
28. Manderla, M.: Eine Methodik zum Regler- und Beobachterentwurf für Mehrgrößensysteme in Deskriptorform. VDI-Verlag, Düsseldorf (2011)
29. Manderla, M., Labisch, D., Konigorski, U.: Dualität von Regler- und Beobachterentwurf für reguläre Deskriptorsysteme. In: Adamy, J., Sawodny, O. (eds.) GMA Fachausschuss 1.40 "Theoretische Verfahren der Regelungstechnik", Workshop, 18–21 Sept 2011, Salzburg, pp. 320–331 (2011)
30. Mehrmann, V., Stykel, T.: Descriptor systems: a general mathematical framework for modelling, simulation and control. At – Automatisierungstechnik **54**, 405–415 (2006)
31. Müller, P.C., Hou, M.: On the observer design for descriptor systems. IEEE Trans. Autom. Control **38**, 1666–1671 (1993)
32. Nijmeijer, H., van der Schaft, A.: Nonlinear Dynamical Control Systems. Springer, New York (1990)
33. Ploix, S., Maquin, D., Ragot, J.: Asymptotic observer for a non linear descriptor system. In: IFAC Symposium on Fault Detection, Supervision and Safety for Technical Processes (SAFEPROCESS'97), pp. 654–658. Hull, Royaume-Uni (1997)

34. Respondek, W.: Right and left invertibility of nonlinear control systems. In: Sussmann, H.J. (ed.) Nonlinear Control and Optimal Control, pp. 133–176. Marcel Dekker, New York (1990)
35. Shields, D.N.: Observer design for discrete nonlinear descriptor systems. In: UKACC International Conference on Control, Exeter, vol. 2, pp. 831–836 (1996)
36. Shields, D.N.: Observer design and detection for nonlinear descriptor systems. Int. J. Control **67**(2), 153–168 (1997)
37. Terrell, W.J.: Local observability of nonlinear differential-algebraic equations (DAEs) from the linearization along a trajectory. IEEE Trans. Autom. Control **46**(12), 1947–1950 (2001)
38. Yang, C., Zhang, Q., Chai, T.: Observer design for a class of nonlinear descriptor systems. In: Proceedings of the 48th IEEE Conference on Decision and Control and 28th Chinese Control Conference, Shanghai, pp. 8232–8237 (2009)
39. Yang, C., Zhang, Q., Zhou, L.: Stability Analysis and Design for Nonlinear Singular Systems, vol. 435. Springer, Berlin/Heidelberg (2013)
40. Zimmer, G., Meier, J.: On observing nonlinear descriptor systems. Syst. Control Lett. **32**(1), 43–48 (1997)

Part II
Simulation

Monitoring Singularities While Integrating DAEs

Diana Estévez Schwarz and René Lamour

Abstract Modern simulation tools for ODEs/DAEs allow a direct input of equations that are solved at the push of a button. However, if the mathematical assumptions that guarantee the correctness of the solution are not given, then no reliable results can be expected. Automatic (or algorithmic) differentiation (AD) opens new possibilities to analyze and solve ODEs/DAEs. In this paper, we outline how the index determination, the computation of consistent initial values, the integration and the diagnosis of singular points can be reliably carried out for DAEs up to index 3. The approach uses the projector based analysis for DAEs employing AD.

Keywords Differential-algebraic equations • Taylor series • Singularity • Automatic differentiation

Mathematics Subject Classification (2010) 65L80 · 65L05 · 41A58

1 Introduction

Differential-algebraic equations (DAEs) are implicit ordinary differential equations in the form

$$f(x'(t), x(t), t) = 0 \quad \text{or, alternatively,} \quad f((d(x(t), t))', x(t), t) = 0,$$

D. Estévez Schwarz
B. Hochschule für Technik Berlin, Luxemburger Str. 10, 13353 Berlin, Germany
e-mail: estevez@beuth-hochschule.de

R. Lamour (✉)
Institut für Mathematik, Humboldt Universität zu Berlin, 10099 Berlin, Germany
e-mail: lamour@math.hu-berlin.de

S. Schöps et al. (eds.), *Progress in Differential-Algebraic Equations*,
Differential-Algebraic Equations Forum, DOI 10.1007/978-3-662-44926-4_4

where the partial Jacobian $f_y(y,x,t)$ does not have full row rank. As a consequence, DAEs may involve not only integration problems, but differentiation problems, too. All index concepts give a kind of measure of the deviations of a DAE from ODEs.

One possibility to handle the differentiation and integration related to DAEs is the employment of automatic (or algorithmic) differentiation (AD). In [4] an overview of the different approaches from literature is given and a new integration method by the projector based analysis is presented for DAEs up to index 2.

The purpose of this paper is twofold. On the one hand, we generalize the results of [4] to index-3 DAEs. On the other hand, we explore and illustrate the possibilities that the projector based analysis of DAEs using AD offers with regard to the diagnosis of singular points. A regular point of a DAE can be characterized by the property that one solution traverses that point only. Singularities appear in various forms. In particular, more than one solution may cross the point or the general solution may have a singularity at that point. All singularities have in common that the tractability index matrix sequence (cf. Sect. 2) is not regular in an open neighborhood of that point (cf. [6, Definition 3.21]).

The basic definitions related to the tractability index concept are introduced in Sect. 2, while the decoupling for index-3 DAEs is specified in Sect. 3. In Sect. 4 we briefly introduce some basics of the algorithmic or automatic differentiation (AD) technique and summarize some notation introduced in [4]. Using AD, the projector based methods allow the index determination, consistent initialization, time integration, and the diagnosis of singular points. These aspects are addressed in Sects. 5 and 6. Our code *daePAD* (DAE solver based on Projectors and AD) was implemented in Matlab using the automatic differentiation toolbox INTLAB [11]. We have compared our calculations with other codes, in which the identification of singularities seems to occur by chance, depending on the treatment of the equations. Several examples are discussed in Sect. 7.

2 Tractability

In this paper we use the projector based analysis of DAEs related to the tractability index concept (cf. [6]). This approach is based on linearization and allows a decoupling of the components and variables with regard to the properties of the DAEs. Some examples that illustrate the matrix sequence from below can be found in Sect. 7. We focus on DAEs with properly stated leading term of the form

$$f((d(x,t))',x,t) = 0. \tag{1}$$

Properly stated means that

$$\ker \frac{\partial f}{\partial y}(y,x,t) \oplus \operatorname{im} \frac{\partial d}{\partial x}(x,t) = \mathbb{R}^n, \tag{2}$$

$\forall y \in \mathscr{D}_y \subseteq \mathbb{R}^n,\ x \in \mathscr{D}_x \subseteq \mathbb{R}^m,\ t \in \mathscr{I}$, and $\operatorname{rank} \frac{\partial f}{\partial y} = \operatorname{rank} \frac{\partial d}{\partial x} =: r_0$ is constant. The projector R that realizes the splitting (2) with $\operatorname{im} R = \operatorname{im} \frac{\partial d}{\partial x}$, $\ker R = \ker \frac{\partial f}{\partial y}$ is assumed to be smooth.

The tractability index bases on a matrix sequence. The matrix functions

$$\begin{aligned} D(x,t) &= d_x(x,t) \in \mathbb{R}^{n\times m},\\ A(y,x,t) &= f_y(y,x,t) \in \mathbb{R}^{m\times n},\\ B(y,x,t) &= f_x(y,x,t) \in \mathbb{R}^{m\times m} \end{aligned} \tag{3}$$

start the matrix sequence

$$\begin{aligned} G_0 &= A\,D, \quad B_0 = B,\\ G_{i+1} &= G_i + B_i\,Q_i,\\ B_{i+1} &= \left(B_i - G_{i+1}\,D^-\,(D\,\Pi_{i+1}\,D^-)'\,D\,\Pi_{i-1}\right)P_i, \end{aligned} \tag{4}$$

where Q_i is an admissible projector function such that $\operatorname{im} Q_i = \ker G_i$, $P_i := I - Q_i$, D^- denotes a reflexive generalized inverse of D with $D^-D = P_0$, and $\Pi_i := P_0 \cdots P_i$. Admissible means that $Q_i(I - \Pi_{i-1}) = 0$ holds.

The first integer μ where the matrix G_μ becomes nonsingular is called the *tractability index*. Recall that the index and the characteristic values $r_i = \operatorname{rank} G_i$ are independent of the choice of the projectors (cf. [6]).

3 Index-3 Decoupling

With regard to a generalization of the results from [4] to the index-3 case, we discuss linear systems in detail. The generalization for the nonlinear index-3 case results straightforward using Newton–Kantorowitsch (see Sect. 5.4).

Let us consider regular linear time-dependent DAEs

$$A(Dx)' + Bx = q$$

with tractability index 3. Such DAEs decouple into (cf. [10, p. 72], [6, formula (2.62) for $\mu = 3$])

$$u' = (D\Pi_2 D^-)'u + D\Pi_2 G_3^{-1}(q - BD^-u), \tag{5}$$

$$v_2 = \Pi_1 Q_2 G_3^{-1}(q - BD^-u), \tag{6}$$

$$v_1 = P_0 Q_1 P_2 G_3^{-1}(q - BD^-u) + P_0 Q_1 Q_2 D^-((Dv_2)' + (D\Pi_2 D^-)'u), \tag{7}$$

$$v_0 = Q_0 P_1 P_2 G_3^{-1}(q - BD^- u) + Q_0(Q_1 + P_1 Q_2)D^-((Dv_2)' + (D\Pi_2 D^-)'u) \\ + Q_0 Q_1 D^-(Dv_1)' + Q_0(P_1 - Q_1)D^-((D\Pi_1 D^-)' - (D\Pi_2 D^-)')Dv_2, \tag{8}$$

where

$$u = D\Pi_2 x, \quad v_2 = \Pi_1 Q_2 x, \quad v_1 = P_0 Q_1 x, \quad v_0 = Q_0 x,$$

and the DAE solution is given by $x = D^- u + v_2 + v_1 + v_0$. Equation (5) represents the inherent ODE, and (6)–(8) assignments to the so-called index-3 component v_2, the index-2 component v_1 and, the algebraic component v_0, respectively.

The decoupling shows that initial values are only possible for the u-component. Therefore, we can formulate an initial condition as

$$D\Pi_2(x(t_0) - \alpha) = 0$$

for a given $\alpha \in \mathbb{R}^m$.

Note that the decoupling (5)–(8) includes the index-2, index-1 and index-0 cases, too. The index-2 case is characterized by $Q_2 \equiv 0$, $P_2 \equiv I$. In the index-1 case, it holds additionally that $Q_1 \equiv 0$, $P_1 \equiv I$. The index-0 case, i.e., an implicit ODE, corresponds to $Q_0 \equiv 0$, $P_0 \equiv I$, $D \equiv I$. The corresponding simplified decoupling for each case can be found in [4].

With regard to the practical computation of the expressions in (6)–(8) we determine all expressions that are needed for more than one component on a previous step. After the determination of u, this involves the solution of the linear system

$$G_3 y = q - BD^- u,$$

the derivative $W := (D\Pi_2 D^-)'$ and, after the computation of v_2, the expression $z := D^-((Dv_2)' + Wu)$. The decoupling simplifies then to

$$v_2 = \Pi_1 Q_2 y, \tag{9}$$

$$v_1 = P_0 Q_1 P_2 y + P_0 Q_1 Q_2 z, \tag{10}$$

$$v_0 = Q_0 P_1 P_2 y + Q_0(Q_1 + P_1 Q_2)z \\ + Q_0 Q_1 D^-(Dv_1)' + Q_0(P_1 - Q_1)D^-((D\Pi_1 D^-)' - W)Dv_2. \tag{11}$$

4 Automatic Differentiation (AD)

According to [4] for the Taylor expansion of a function x at t_0 we will use the notation

$$x^{[0:K]} = x(t^{[0:K]}) = \begin{bmatrix} x(t_0) & x'(t_0) & \frac{x''(t_0)}{2} & \dots & \frac{x^{(K)}(t_0)}{K!} \end{bmatrix},$$

whereas for the independent variable $t \in I$ with $K \geq 1$, which is initialized with the value t_0, we have

$$t^{[0:K]} := [t_0 \quad 1 \quad 0 \quad \dots \quad 0].$$

To realize the differentiation and integration of Taylor expansions we will use the shifts and the notation introduced in [4]. In particular, for an arbitrary Taylor expansion

$$z^{[0:K]} =: [c_0, c_1, \dots, c_K], \quad c_j \in \mathbb{R}^m, \quad 0 \leq j \leq K,$$

and a constant vector $C \in \mathbb{R}^m$ we define

$$\left[(z^{[0:K]})', C\right] := \left[c_1, 2c_2, \dots, jc_j, \dots, Kc_K, C\right],$$

$$\left[C, \int z^{[0:K]}\right] := \left[C, c_0, \frac{c_1}{2}, \dots, \frac{c_j}{j+1}, \dots, \frac{c_{K-1}}{K}\right]$$

and introduce the linear mapping

$$S_d(z^{[0:K]}) := \left[(z^{[0:K]})', 0\right].$$

In order to cope with the difficulties occurring with these shifts, we will use the notation $x^{[0:K-j,*]}$ or $x^{[*,j:K]}$ for $j \in \mathbb{N}$ to emphasize the fact that the last $K - j + 1, \dots, K$ or the first $0, \dots, j-1$ coefficients may be insignificant or incorrect. Further, in order to compare only a part of the terms of two Taylor expansions, we will use an underset equality specifying the considered terms. A more detailed introduction to the notation we use can be found in [4].

5 Consistent Initialization and Integration with AD

We will first consider the IVP for linear index-μ DAEs

$$A(Dx)' + Bx = q, \quad D\Pi_{\mu-1}(x(t_0) - \alpha) = 0.$$

Dealing with AD, we compute $A^{[0:K]} := A(t^{[0:K]})$, $D^{[0:K]} := D(t^{[0:K]})$, $B^{[0:K]} := B(t^{[0:K]})$, and $q^{[0:K]} := q(t^{[0:K]})$, and consider the operator equation

$$A^{[0:K]} \cdot \left[\left(D^{[0:K]} x^{[0:K-1,*]} \right)', 0 \right] + B^{[0:K]} \cdot x^{[0:K-1,*]} \underset{[0:K-1]}{=} q^{[0,K]}, \tag{12}$$

$$D\Pi_{\mu-1}^{[0]} (x^{[0]} - \alpha) = 0. \tag{13}$$

The procedure to solve these equations consists of three steps: index determination and corresponding specification of $D\Pi_{\mu-1}^{[0]}$ (cf. Sect. 5.1), consistent initialization (cf. Sect. 5.2), integration (cf. Sect. 5.3). Note that the index determination and the consistent initialization can be used stand-alone, for diagnosis and/or as previous step if other integration methods are used.

With regard to the generalization of the approach for nonlinear DAEs, in this paper we briefly outline how the approach from [4] for index-2 DAEs can be adapted for the index-3 case in Sect. 5.4. The basic idea is to use the Newton–Kantorowitsch method.

5.1 *Index Determination*

The index determination requires the computation of the matrix sequence (4) starting with the Taylor coefficients of the matrices A, D and B. The nullspace projectors onto $\ker G_i$, $i \geq 0$ are chosen widely orthogonal (cf. [6, Chapter 7]) and computed by Householder decomposition. The necessary derivatives are realized by the corresponding shifts (cf. Sect. 4). Since the definition of G_i, $i \geq 2$ involves a derivative in B_{i-1}, we may only obtain $G_i^{[0:K-i+1,*]}$ and, accordingly, $Q_i^{[0:K-i+1,*]}$.

Hence, for $K \geq 2$, a reliable index characterization up to index 3 is possible. If the matrix sequence terminates with a nonsingular matrix G_μ, all needed matrices and, in particular, the projectors are available. For $\mu \leq 3$, we can apply the decomposition (5) and (9)–(11) with regard to the following aspects.

5.2 *Consistent Initialisation*

To compute consistent initial values, the operator equation (12)–(13) is solved. For index-2 DAEs, the idea of our approach is exposed in [4]. It consists in decoupling the linear DAE in order, to apply the Taylor series method to the inherent ODE and to assign, afterwards, the corresponding values to the remaining components.

Here, we focus on the index-3 case, i.e., we consider $D\Pi_2$ in (13). Considering $G_3^{[0:K-2,*]}$, the decoupling (5) and (9)–(11) directly leads to the following steps at a time point t_j assuming $K \geq 4$:

(i) Compute $q^{[0:K]} := q\left(t^{[0:K]}\right)$ and use $D^{[0:K]}$, $B^{[0:K]}$, $(D^-)^{[0:K]}$, $Q_0^{[0:K]}$, $P_0^{[0:K]}$, $Q_1^{[0:K]}$, $P_1^{[0:K]}$, $\Pi_1^{[0:K]}$, $Q_2^{[0:K-1,*]}$, $P_2^{[0:K-1,*]}$, $\Pi_2^{[0:K-1,*]}$, and $G_3^{[0:K-2,*]}$ from the matrix sequence.
(ii) Set $u^{[0]} = (D\Pi_2)^{[0]}\alpha$ and $u^{[0,*]} = [u^{[0]}, 0, \ldots, 0]$.
(iii) For $K_e = 0, \ldots, K-2$.

(a) Solve

$$G_3^{[0:K-2,*]} \cdot y^{[0:K_e,*]} = q^{[0:K]} - (BD^-)^{[0:K]} u^{[0:K_e,*]}.$$

(b) Set

$$u_p^{[0:K_e,*]} := S_d\left((D\Pi_2 D^-)^{[0:K-1,*]}\right) u^{[0:K_e,*]} + (D\Pi_2)^{[0:K-1,*]} y^{[0:K_e,*]}.$$

(c) Set

$$u^{[0:K_e+1,*]} := \left[u^{[0]}, \int u_p^{[0:K_e,*]}\right].$$

Note that, in the end, we obtain $u_p^{[0:K-2;*]}$, due to the shift and the fact that, for G_3, the last two terms of the Taylor expansion may not be correct. However, we finally obtain $u^{[0:K-1]}$.

(iv) Solve

$$G_3^{[0:K-2,*]} \cdot y^{[0:K-2,*]} = (q - BD^- u)^{[0:K-1]},$$

compute

$$W^{[0:K-2,*]} := S_d\left((D\Pi_2 D^-)^{[0:K-1,*]}\right),$$

and set

$$v_2^{[0:K-2,*]} = (\Pi_1 Q_2)^{[0:K-1,*]} y^{[0:K-2,*]},$$

$$z^{[0:K-3,*]} := (D^-)^{[0:K]}\left(S_d\left((Dv_2)^{[0:K-2,*]}\right) + W^{[0:K-2,*]} u^{[0:K-1,*]}\right)$$

$$v_1^{[0:K-3,*]} = (P_0 Q_1 P_2)^{[0:K-1,*]} y^{[0:K-2,*]} + (P_0 Q_1 Q_2)^{[0:K-1,*]} z^{[0:K-3,*]},$$

$$
\begin{aligned}
v_0^{[0:K-4,*]} =& (Q_0P_1P_2)^{[0:K-1,*]}y^{[0:K-2,*]} + (Q_0(Q_1+P_1Q_2))^{[0:K-1,*]}z^{[0:K-3,*]}\\
&+ (Q_0Q_1D^-)^{[0:K]}S_d\left((Dv_1)^{[0:K-3,*]}\right)\\
&+ (Q_0(P_1-Q_1)D^-)^{[0:K]}\left(S_d\left((D\Pi_1D^-)^{[0:K]}\right) - W^{[0:K-2,*]}\right)\\
&\ (Dv_2)^{[0:K-2,*]}.
\end{aligned}
$$

Observe that we obtain $v_0^{[0:K-4,*]}$ in the worst case.

(v) Set $x^{[0:K-4,*]} = (D^-u)^{[0:K]} + v_2^{[0:K-2,*]} + v_1^{[0:K-3,*]} + v_0^{[0:K-4,*]}$.

Note that, if $(D\,\Pi_1\,D^-)^{(K+1)} \equiv 0$, then we deal with $G_2^{[0:K]}$ and $G_3^{[0:K-1,*]}$, indeed. If it further holds that $(D\,\Pi_2\,D^-)^{(K+1)} \equiv 0$, then we obtain $G_3^{[0:K]}$. This explains why we compute $y^{[0:K-2,*]}$ anew in Step (iv).

Hence, depending on the structural properties, this step may allow the computation of $y^{[0:K]}$, consequently leading to

$$
x^{[0:K-2,*]} = (D^-u)^{[0:K]} + v_2^{[0:K,*]} + v_1^{[0:K-1,*]} + v_0^{[0:K-2,*]}.
$$

Moreover, if $(Dv_1)^{(K+1)} \equiv 0$ and $(Dv_2)^{(K+1)} \equiv 0$ holds additionally, then we even obtain $x^{[0:K]} = (D^-u)^{[0:K]} + v_2^{[0:K,*]} + v_1^{[0:K]} + v_0^{[0:K]}$.

Above we assumed $K \geq 4$ in general. In fact, if we are only interested in the computation of a consistent initialization, $K \geq 4$ will be sufficient. However, for a reasonable integration Taylor step we need $K > 4$ instead (cf. Sect. 5.3).

With regard to DAEs with index $\mu > 3$ we observe that, by definition, the decoupling from [6] would lead to $G_\mu^{[0:K-(\mu-1),*]}$ in general, and consequently to $v_{\mu-1}^{[0:K-(\mu-1),*]}$, $v_{\mu-2}^{[0:K-\mu,*]}$, $\ldots v_0^{[0:K-2(\mu-1),*]}$. Hence, for consistent initialization we have to choose $K \geq 2(\mu-1)$ and for the Taylor series method $K > 2(\mu-1)$.

5.3 *Integrating IVPs*

If the Taylor expansion

$$
x^{[0:\tilde{K}]} = \left[c_0, \ldots, c_{\tilde{K}}\right]
$$

for $\tilde{K} \leq K$ is known at t_j, then the computation of an initial guess α for the next time-step results from

$$
\alpha = x_{j+1} := \sum_{l=0}^{\tilde{K}} c_l h_j^l \approx x(t_j + h_j)
$$

for a given h_j that can be computed by a time-step control strategy [4]. Afterwards, the consistent initialization at $t_{j+1} = t_j + h_j$ is computed and the procedure is repeated.

Note that due to its explicit nature, the Taylor series method is not well suited in case of stiffness.

5.4 *Nonlinear DAEs*

In the nonlinear case we consider (1)

$$f((d(x,t))', x, t) = 0$$

at a fixed t_j as an operator equation

$$F(x) = 0 \tag{14}$$

with $F(x) := f((d(x,\cdot))', x, \cdot)$. According to [6, Section 3.9] for the theoretical framework, we consider $d(x,t) = D(t)x$. Then, the derivative of F at a fixed x_* reads

$$F'(x_*)x = A_*(Dx)' + B_*x.$$

The formal application of the Newton–Kantorowitsch method means the successive solution of the linearization of (14)

$$F'(x_i)\delta x_i = -F(x_i), \quad x_{i+1} = x_i + \delta x_i. \tag{15}$$

The known properties of the operator F, especially in the higher index case, do not allow the direct application of the Newton–Kantorowitsch method (cf. [6, Section 3.9]) in general. However, these results focus on minimal smoothness conditions. In contrast, since we aim at an approach based on Taylor expansions, here we have to assume higher smoothness of all involved operators and solutions anyway.

Therefore we suppose that the solution of the DAEs can be approximated by truncated Taylor series. Consequently, we are looking for the coefficients c_i, $0 \leq i \leq K$, of the Taylor expansion of the solution x of (14) and consider

$$f\left(S_d\left(d\left(x^{[0:K-1,*]}, t^{[0:K]}\right)\right), x^{[0:K-1,*]}, t^{[0:K]}\right) \underset{[0:K-1]}{=} 0. \tag{16}$$

The linearization is related to (12)–(13) with

$$\begin{aligned}
D^{[0:K,*]} &= D\left(x^{[0:K,*]}, t^{[0:K]}\right), \\
A^{[0:K-1,*]} &= A(S_d\left(d\left(x^{[0:K,*]}, t^{[0:K]}\right), x^{[0:K]}, t^{[0:K]}\right)), \\
B^{[0:K-1,*]} &= B(S_d\left(d\left(x^{[0:K,*]}, t^{[0:K]}\right)\right), x^{[0:K]}, t^{[0:K]}), \text{ and} \\
q^{[0:K-1,*]} &= -f\left(S_d\left(d\left(x^{[0:K,*]}, t^{[0:K]}\right)\right), x^{[0:K,*]}, t^{[0:K]}\right).
\end{aligned}$$

For a nonlinear DAE we can therefore apply the algorithm of Sect. 5.2 with these matrices A, D, B and this right-hand side q successively. Analogously to [4], δx_i is computed step by step considering δu_i, δv_{2i}, δv_{1i}, δv_{0i}. Note finally that, in the index-μ case, we start the Newton–Kantorowitsch loop with $x_0^{[0]} = \alpha$ and set

$$\delta u_i^{[0]} = (D\Pi_{\mu-1})_i^{[0]}(\alpha - x_i^{[0]})$$

in order to cope with the fact that the matrix sequence itself and, hence, also the projectors depend on the solution.

6 Diagnosis of Singular Points

The tractability index definition (cf. Sect. 2) requires constant ranks $r_i = \operatorname{rank} G_i$. If one of these ranks is not constant at an environment of a point, we call it a singular point. First investigations to classify singular points are made in [1, 7, 8]. At a singular point, we cannot predict the behavior of the solution, since corresponding theoretical results are missing. Hence, the identification seems to be a crucial concern to ensure the reliability of numerical results.

In the linear case, the matrices A, D and B may depend on time and t_* will be identified as a critical point if there is no open interval containing t_* such that the ranks r_i remain constant on it. For a more detailed definition of critical points and a regularity interval we refer to [6].

In the general nonlinear case, the situation is much more complicated. The matrices A, D and B of (3) may depend on time and the state y and x. Related to (1) the argument y is connected with $(d(x,t))'$. That means that the rank of G_0 may depend on the state x and its derivative. Because of the time derivative computing B_i, $i > 0$, in the worst case the rank of G_i may depend on higher derivatives of x. The special structure of

$$G_{i+1} = G_i + B_i Q_i = (G_i + B_{i-1}P_{i-1}Q_i)\underbrace{(I - P_i D^-(D\Pi_i D^-)' D\Pi_{i-1}Q_i)}_{\textit{nonsingular}}$$

yields that the rank of G_{i+1} (and therefore also the index of the DAE) depends at most on an i-th derivative for $i > 0$. G_0 and G_1 depend at most on the first derivative. See also Subsection 3.2 of the chapter on Nonlinear DAEs of [6].

According to [6] on a so-called regularity region $\mathcal{G} \in \mathcal{D}_x \times \mathcal{I}$, the characteristic values r_i are constant. In *daePAD*, we precisely check this condition. In the following examples we show that, with *daePAD*, the identification of singular points becomes possible.

7 Examples

All integration methods indirectly identify those critical points that correspond to singular points of the general solution, in particular poles. Normally, the step-size control is decreased to a minimum and the integration stops. In contrast, if multiple solutions pass through a point, no warning is given in general. Hence, we will focus on these cases and present a selection of examples to illustrate our approach. For comparison, we tested some of these examples also with Matlab (Version 2012b), the demo version of Dymola (Version 2013) and OpenModelica (Version 1.8.1).

At this point we also want to emphasize that the computation of consistent initial values for index-3 DAEs is, in general, a numerical challenge. Our algorithm was successful for many tested examples, especially for Example 7.

7.1 Linear DAEs

Several solvers of DAEs that use AD are based on the structural index concept. However, even for linear DAEs with constant coefficients the structural index may differ from the expected one, according to different definitions. Considering the well-known example from [9], we illustrate that an approach based on the tractability index concept does not have this drawback.

Example 1 (Comparison with structural index) The DAE

$$-x_1' + x_2' - x_3 = 0 \tag{17}$$

$$x_1' - x_2' + x_2 = 0 \tag{18}$$

$$-x_1 = \cos(t) \tag{19}$$

can be rewritten in the form

$$A(Dx)' + Bx = q$$

for

$$A = \begin{pmatrix} -1 \\ 1 \\ 0 \end{pmatrix}, \quad D = \begin{pmatrix} 1 & -1 & 0 \end{pmatrix}, \quad B = \begin{pmatrix} 0 & 0 & -1 \\ 0 & 1 & 0 \\ -1 & 0 & 0 \end{pmatrix}, \quad \text{and} \quad q = \begin{pmatrix} 0 \\ 0 \\ \cos(t) \end{pmatrix}.$$

The obtained matrix sequence reads

$$G_0 = \begin{pmatrix} -1 & 1 & 0 \\ 1 & -1 & 0 \\ 0 & 0 & 0 \end{pmatrix}, \quad Q_0 = \begin{pmatrix} 0 & 1 & 0 \\ 0 & 1 & 0 \\ 0 & 0 & 1 \end{pmatrix}, \quad G_1 = \begin{pmatrix} -1 & 1 & -1 \\ 1 & 0 & 0 \\ 0 & -1 & 0 \end{pmatrix}.$$

This well known example has differentiation and tractability index 1, but it does not have structural index 1 (cf. [9]).

If we fix the initial values by

$$D\left(x(t_0) - \begin{pmatrix} -1 \\ 0 \\ 1 \end{pmatrix}\right) = 0,$$

then we obtain the solution

$$x(t) = \begin{pmatrix} -\cos(t) \\ -\frac{1}{2}(-e^t + \cos(t) + \sin(t)) \\ -\frac{1}{2}(-e^t + \cos(t) + \sin(t)) \end{pmatrix}.$$

Starting with $\alpha = (-1 \;\; 0 \;\; 1)^T$ at $t_0 = 0$, the consistent initial values are $x(t_0) = (-1 \;\; 0 \;\; 0)^T$ and $x'(t_0) = (0 \;\; 0 \;\; 0)^T$. Solving this IVP with *daePAD* for $K = 7$, the computed 8 Taylor coefficients coincide with the exact solution up to the penultimate digit. This accuracy was expected for constant coefficient examples.

The above example illustrates the advantage of an analysis based on the matrix sequence even in the index-1 case. In the higher index case, the constellation that leads to the failure of the structural analysis for the index determination are possibly not that obvious.

For linear DAEs with variable coefficients

$$A(t)\,(D(t)x(t))' + B(t)x(t) = q(t)$$

the rank r_i of the sequence $G_i(t)$ may depend on time, too. In general, singular points may appear and should be identified along the numerical solution. In this context, a pure structural analysis fails by construction, since no numerical values are considered. In contrast, the analysis based on the matrix sequence that is computed using AD is pretty reliable.

In order to illustrate our approach, we start with the following simple scalar ODE.

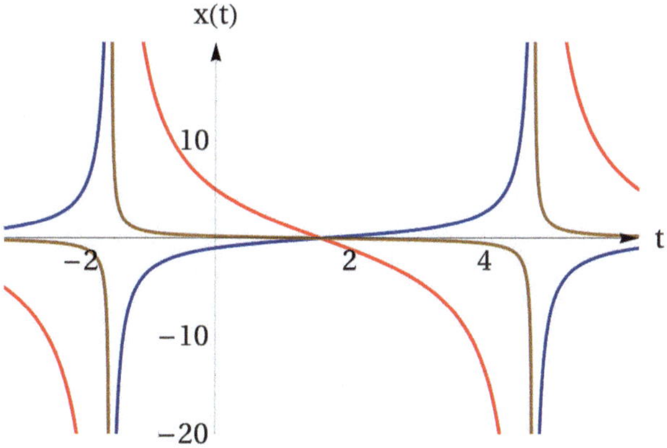

Fig. 1 Solutions of ODE (20) for C = −1, 1/5, 5

Example 2 (ODE with singularity) We consider the linear time dependent ODE

$$\cos(t)\, x' + x = 0, \tag{20}$$

leading to $G_0 = \cos(t)$. Consequently, for $t \neq \frac{\pi}{2} + k\pi$, $k \in \mathbb{Z}$, the index is 0, i.e., (20) is an ODE and we can prescribe an initial value x_0. However, r_0 is not constant. At $t_k = \frac{\pi}{2} + k\pi$ we obtain $G_0 = 0$. In fact, x_k is determined by $x_k = 0$ without any dynamical considerations. Consequently, we cannot prescribe an initial value. More precisely, the solutions of (20) are

$$x(t) = C\frac{\cos(t)}{1+\sin(t)}.$$

Figure 1 illustrates that all solutions intersect at $t_k = \frac{\pi}{2} + k \cdot 2\pi$.

In practice, near $t = \pi/2$ all numerical solvers may easily "jump" from one solution to another, since all solutions lie very close to each other. We illustrate this behavior plotting the difference between the values of the solution $\frac{\cos(t)}{1+\sin(t)}$ and numerical solutions. We tested the initial value problem (20) with $x_0 = x(0) = 1$ using the Matlab solvers ode15s and ode23t with default tolerances. As can be seen in Fig. 2, this difference considerably increases after the singularity. Additionally, time-step control forces very small stepsizes that, in a certain sense, give hint to the numerical difficulties that arise at $\pi/2$.

For the sake of simplicity, the following three illustrative examples are semi-explicit DAEs, i.e., we can set

$$A = \begin{pmatrix} I \\ 0 \end{pmatrix}, \quad D = \begin{pmatrix} I & 0 \end{pmatrix}, \quad G_0 = \begin{pmatrix} I & 0 \\ 0 & 0 \end{pmatrix}, \quad Q_0 = \begin{pmatrix} 0 & 0 \\ 0 & I \end{pmatrix}, \tag{21}$$

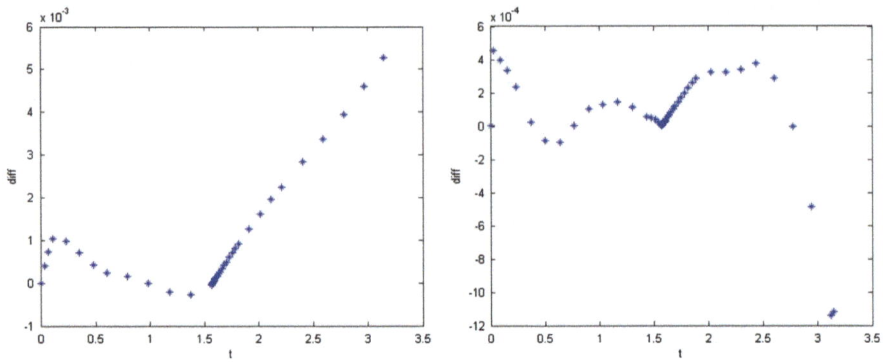

Fig. 2 Difference *diff* between $\frac{\cos(t)}{1+\sin(t)}$ and the numerical solution of (20) with $x_0 = 1$ using ode15s (*left*) and ode23t (*right*)

for the correspondingly sized identity matrix I. Consequently, the time-dependency of the matrix sequence is due to $B(t)$ only. However, we want to emphasize that this structure is not required for *daePAD*, as shown with the Examples 1 and 2.

Example 3 (Index-1 DAE with singularity) If we reformulate the ODE (20) as a semi-explicit index-1 DAE, analogous results will be obtained. The equations

$$x_1' - x_2 = 0, \tag{22}$$

$$x_1 + \cos(t)\, x_2 = 0, \tag{23}$$

lead to

$$B(t) = \begin{pmatrix} 0 & -1 \\ 1 & \cos(t) \end{pmatrix}, \quad G_1 = \begin{pmatrix} 1 & -1 \\ 0 & \cos(t) \end{pmatrix}.$$

For $t \neq \frac{\pi}{2} + k\pi$, $k \in \mathbb{Z}$, the index is 1 and we can prescribe an initial value for x_1, i.e., for Dx. At $t_k = \frac{\pi}{2} + k\pi$ the rank of G_1 changes and the behavior is analogous as before.

Although the computation of a pretty precise matrix sequence is possible thanks to AD, the identification of the time-point t^* at which the rank change occurs during the integration is not trivial in general. In fact, a numerical rank decision is a delicate problem in general. Consequently, we use a Householder decomposition.

In *daePAD*, we monitor the condition of G_μ, the sign of the determinant of G_μ and the used minimal pivot element when decomposing G_μ. For G_i, $i < \mu$, the characteristic values r_i are determined by comparing the values of the pivot elements.

In case a singularity is detected, our integrations stops by default. From our point of view, this is the only reasonable behavior, since no reliable results can be expected thereafter. Nevertheless, we can optionally force the integrator to continue for testing purposes.

Due to the behavior of $\cos(t)$ near $t = \pi/2$, for the last two examples the most reliable indicator was the change of the sign of the determinant G_μ. For the initial value problems (20) with $x_0 = 1$ and (22)–(23) with $x_0 = [1, -1]$ on $[0, \pi]$, our solver reports a rank change and stops at $\pi/2$.

However, if the sign of the determinant of G_μ does not change (cf. Example 4), then this indicator does not work. In fact, a search strategy looking for e.g. a minimal condition seems to be more reasonable, although the condition of the matrix sequence may change depending on the chosen projectors. Of course, if by chance the integration step jumps exactly to t^*, then the singularity is correctly identified.

Example 4 (Index-1 DAE with singularity and $\det G_1 \geq 0$*)* For the DAE

$$x_1' - x_2 = 0,$$
$$-tx_1 + t^2 x_2 = 0$$

we obtain

$$B(t) = \begin{pmatrix} 0 & -1 \\ -t & t^2 \end{pmatrix}, \quad G_1 = \begin{pmatrix} 1 & -1 \\ 0 & t^2 \end{pmatrix}.$$

The DAE has index 1 for $t \neq 0$ and $\det G_1 = t^2$ does not change its sign at $t^* = 0$. Since the general solution is

$$x = C \begin{pmatrix} t \\ 1 \end{pmatrix},$$

big timesteps can be made in general and the singularity at t^* may not be detected. Of course, general techniques from root finding algorithms could be used in such cases. Since our focus is to monitor the singularities while integrating an IVP problem such strategies are not employed.

Let us finally consider an index-2 DAE with a clear but slightly more complex structure and a given analytical solution.

Example 5 (Index-2 DAE with singularity) We consider the system

$$\begin{aligned} x_1' + x_5 &= 0, \\ x_2' + x_5 &= 0, \\ x_3' + x_1 &= 0, \\ x_4' + x_2 + \sin(t)\, x_5 &= 0, \\ x_3 + x_4 &= 0. \end{aligned} \tag{24}$$

With the widely orthogonal projectors the obtained matrix sequence reads

$$G_1 = G_0 + BQ_0 = \begin{pmatrix} 1 & 0 & 0 & 0 & 0 \\ 0 & 1 & 0 & 0 & 0 \\ 0 & 0 & 1 & 0 & 0 \\ 0 & 0 & 0 & 1 & \sin(t) \\ 0 & 0 & 0 & 0 & 0 \end{pmatrix}, \quad Q_1 = \begin{pmatrix} a(t) & a(t) & 0 & b(t) & 0 \\ a(t) & a(t) & 0 & b(t) & 0 \\ 0 & 0 & 0 & 0 & 0 \\ b(t) & b(t) & 0 & b(t)\sin(t) & 0 \\ -a(t) & -a(t) & 0 & -b(t) & 0 \end{pmatrix}$$

with $a(t) = \dfrac{1}{2+\sin^2(t)}$ and $b(t) = a(t)\sin(t)$,

$$G_2 = \begin{pmatrix} 1 & 0 & 0 & 0 & 1 \\ 0 & 1 & 0 & 0 & 1 \\ a(t) & a(t) & 1 & b(t) & 0 \\ a(t)(1+\cos(t)) & a(t)(1+\cos(t)) & 0 & 1-b(t)(1+\cos(t)) & \sin(t) \\ -b(t) & -b(t) & 0 & -b(t)\sin(t) & 0 \end{pmatrix}.$$

The index of (24) is 2 if $\det G_2 = \sin(t) \neq 0$, i.e., for $t \neq k\pi$. The general solution of (24) in the interval $(-\pi, \pi)$ is given by

$$\begin{aligned} x_1(t) &= c_1 - \frac{c_1 + c_2}{1+\cos(t)}, \\ x_2(t) &= c_2 - \frac{c_1 + c_2}{1+\cos(t)}, \\ x_3(t) &= c_3 - c_1 t + (c_1 + c_2)\tan\frac{t}{2}, \\ x_4(t) &= -c_3 + c_1 t - (c_1 + c_2)\tan\frac{t}{2}, \\ x_5(t) &= (c_1 + c_2)\frac{\tan\frac{t}{2}}{1+\cos(t)}. \end{aligned} \tag{25}$$

Note that in the special case that $c_1 + c_2 = 0$ holds, Eq. (25) corresponds to

$$\begin{aligned} x_1(t) &= c_1, \\ x_2(t) &= -c_1, \\ x_3(t) &= c_3 - c_1 t, \\ x_4(t) &= -c_3 + c_1 t, \\ x_5(t) &= 0, \end{aligned} \tag{26}$$

which is defined for all t.

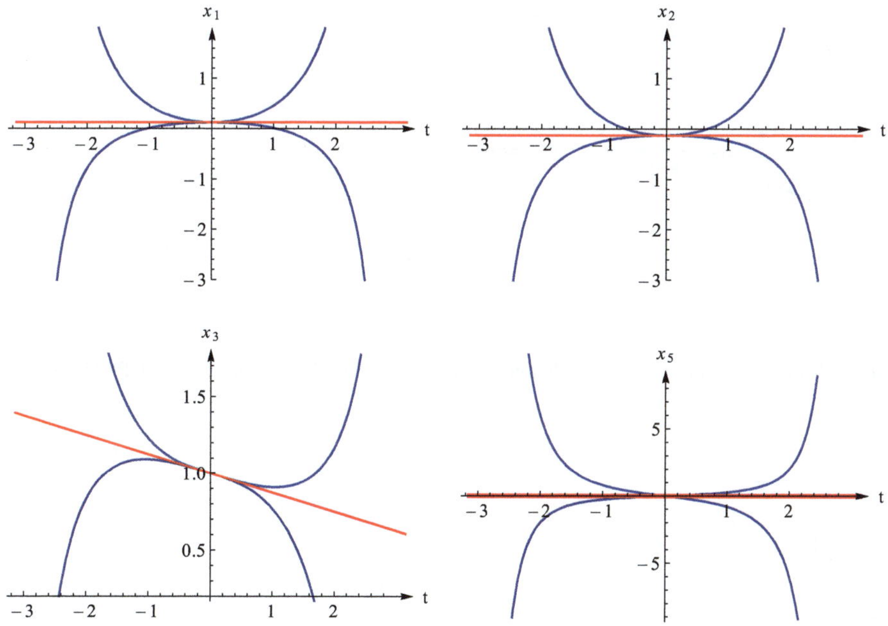

Fig. 3 Solutions of DAE (25) with singularity at $t = 0$

Observe further that the solution (25) has not any singularity at $t = 0$, but that the matrix sequence is singular at that point. In fact, all solutions with $c_1 - c_2 = \alpha$, $c_3 = \beta$ for fixed α and β intersect at $t = 0$ in $[\alpha/2, -\alpha/2, \beta, -\beta, 0]$.

For $c_1 - c_2 = 1/4$ and $c_3 = 1$ at $t = 0$ all solutions pass through $[1/8, -1/8, 1, -1, 0]$ (see Fig. 3). We solved the IVP of the DAE (24) with the initial condition $D\Pi_1\left(x(-\frac{\pi}{2}) - \alpha\right) = 0$ with $\alpha = (-0.25\ -0.5\ 1\ -1\ 1)^T$. These values lead to $c_1 = 1/2, c_2 = 1/4$, and $c_3 = 1$ and the exact consistent initial values at $t_0 = -\frac{\pi}{2}$ are

$$(-c2,\ -c1,\ c3 - c1 \cdot t_0 - (c1 + c2),\ -(c3 - c1 \cdot t_0 - (c1 + c2)),\ -(c1 + c2))^T,$$

whereas the exact values for the third and fourth component are $x_3(t_0) = -x_4(t_0) = 1.035398163397448$. The consistent initial value $x(t_0)$ computed by *daePAD* reads

$$(-2.49\ldots 97\text{e-}01, -5.\text{e-}01, 1.035398163397448, -1.035398163397448, -7.5\text{e-}01)^T.$$

For these initial values, Dymola using DASSL or LSODAR and *daePAD* compute similar numerical solutions (see left graphs of Figs. 4 and 5). Using the solvers DOPRI45 and RADAU IIa – order 5 stiff Dymola computes a solution which behaves remarkably (for DOPRI45 see right-hand side of Fig. 4). However,

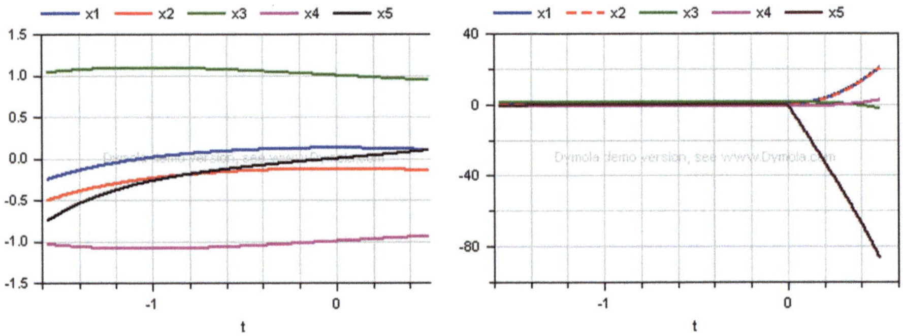

Fig. 4 Numerical solutions of (25) obtained by Dymola using DASSL (*left*) and DOPRI45 (*right*)

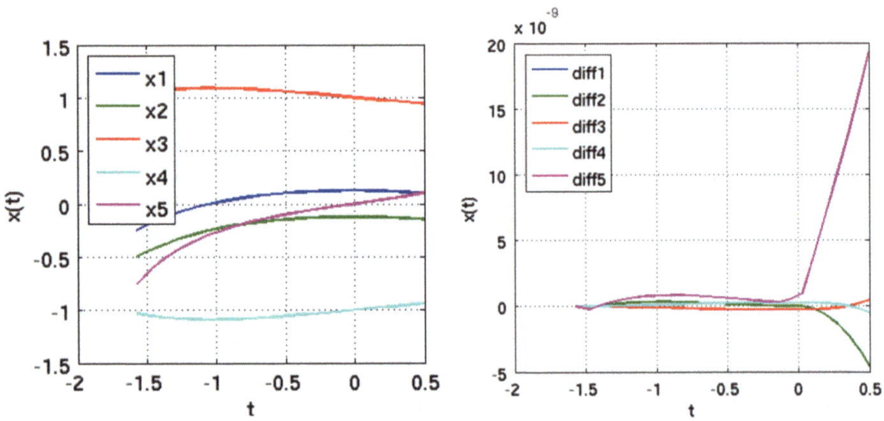

Fig. 5 Numerical solution obtained with *daePAD* forcing the integration after the singular point (*left*) and corresponding difference between this solution and the solution (25) for $c_1 = 1/2, c_2 = 1/4$ and $c_3 = 1$ (*right*)

this behavior may also be traced back to the internal algorithm used to reformulate DAEs.

Note that the solutions cross the regularity region border $t = 0$. Hence, by default *daePAD* terminates at that point. In case it is forced to continue the integration, it will always report the detected singularities. Next to $t = 0$, the indicators used for the detection of a singularity are shown in Fig. 6. In contrast, Dymola delivers a solution without any warning.

Recall further that, although solutions may look reasonable at first glance, after a singular point they may not correspond to the former solution anymore. To highlight this behavior, we represent the difference between the obtained numerical solutions and the solution (25) for $c_1 = 1/2, c_2 = 1/4$ and $c_3 = 1$ in Fig. 7 and the right graph of Fig. 5.

t	det G_2	cond G_2	min pivot	index
-1.570796e+00	-1.000000e+00	4.476344e+00	5.388159e-01	2
-1.471488e+00	-9.950729e-01	4.562153e+00	5.204151e-01	2
⋮	⋮	⋮	⋮	⋮
-1.347154e-01	-1.343083e-01	4.495943e+01	6.360231e-02	2
-5.331210e-02	-5.328685e-02	1.156359e+02	2.599708e-02	2
2.758192e-02	2.757843e-02	2.269872e+02	1.398874e-02	2
1.083718e-01	1.081598e-01	5.853485e+01	5.760912e-02	2
⋮	⋮	⋮	⋮	⋮
5.000000e-01	4.794255e-01	1.328674e+01	3.310631e-01	2

Fig. 6 Indicators observed for monitoring a change of r_2 for Example 5. The change of the sign of det G_2 safely detects the singularity. The characteristic values r_0 and r_1 are constant

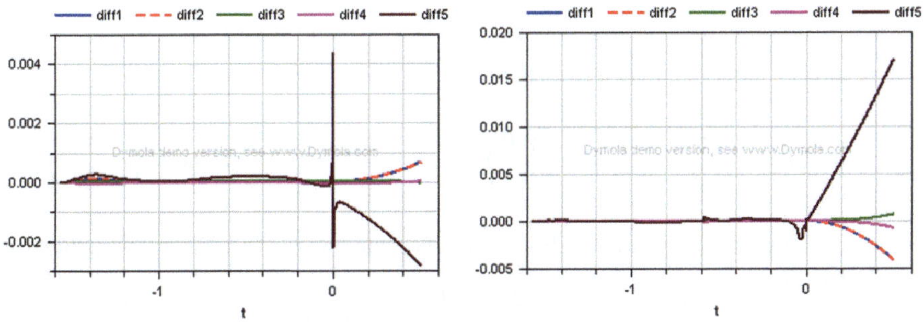

Fig. 7 Difference between the solution (25) for $c_1 = 1/2, c_2 = 1/4$ and $c_3 = 1$ and the numerical solution obtained by Dymola using DASSL (*left*) and LSODAR (*right*)

7.2 Nonlinear Examples

The following simple example illustrates the necessity to identify singular points during the numerical integration. For the considered linear examples from above, it was relatively obvious that the rank of a matrix G_i changes if a key entry of the matrix B becomes zero at a certain time point. In the nonlinear case, no clue about the critical time points is given in general.

Example 6 (Nonlinear index-1 DAE with singularity) We consider the nonlinear index-1 example

$$\begin{aligned} x_1' - x_2 &= 0, \\ x_1^2 + x_2^2 - 1 &= 0 \end{aligned} \tag{27}$$

with

$$A = \begin{pmatrix} 1 \\ 0 \end{pmatrix}, \quad D = \begin{pmatrix} 1 & 0 \end{pmatrix} \text{ and } B = \begin{pmatrix} 0 & -1 \\ 2x_1 & 2x_2 \end{pmatrix}.$$

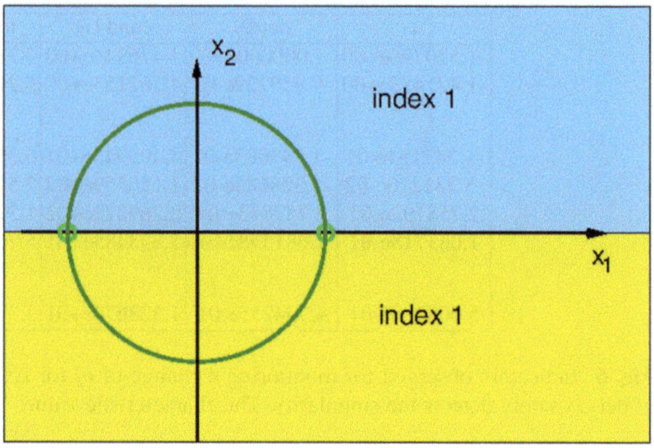

Fig. 8 Regularity regions $\mathscr{G}_1$ (*blue*), $\mathscr{G}_2$ (*yellow*) and solutions of Example 6 (*green*)

The resulting sequence matrices (4) are

$$G_0 = AD = \begin{pmatrix} 1 & 0 \\ 0 & 0 \end{pmatrix}, \quad Q_0 = \begin{pmatrix} 0 & 0 \\ 0 & 1 \end{pmatrix}, \quad \text{and} \quad G_1 = G_0 + BQ_0 = \begin{pmatrix} 1 & -1 \\ 0 & 2x_2 \end{pmatrix}.$$

For $x_2 \neq 0$ the DAE has index 1. For $x_2 = 0$ we can choose the admissible projector $Q_1 = \begin{pmatrix} 1 & 0 \\ 1 & 0 \end{pmatrix}$, obtaining $G_2 = G_1 + B_0 P_0 Q_1 = \begin{pmatrix} 1 & -1 \\ 2x_1 & 0 \end{pmatrix}$, i.e., for $x_1 \neq 0, x_2 = 0$ the linearization of (27) has index 2.

The solutions are $x = \begin{pmatrix} \sin(t+\varphi) \\ \cos(t+\varphi) \end{pmatrix}$ for a constant phase shift $\varphi \in \mathbb{R}$ and $x = \begin{pmatrix} \pm 1 \\ 0 \end{pmatrix}$.

The regularity regions (see Sect. 6) are

$$\mathscr{G}_1 = \{x |\, x_2 > 0\} \quad \text{and} \quad \mathscr{G}_2 = \{x |\, x_2 < 0\},$$

represented in Fig. 8.

For $x_2 \neq 0$ the initial condition reads $D(x(0) - \alpha) = x_1(0) - \alpha_1 = 0$ for an admissible $\alpha = (\alpha_1, \alpha_2)$, i.e. an α with $|\alpha_1| \leq 1$.

We solved this simple example for $[0, 2\pi]$ with Matlab, OpenModelica and Dymola and observed tree different behaviors, which are represented in Fig. 9:

(i) The integration stops around $t = \frac{\pi}{2}$, since the step size is below the smallest value allowed. This result is obtained for most of the integrators using the default tolerances.
(ii) The results correspond to the trigonometric functions on $[0, 2\pi]$. This behavior is obtained with some OpenModelica solvers with default tolerances and with Matlab (ode15s, ode23t) if we use crude tolerances (RelTol=AbsTol=1e-3).

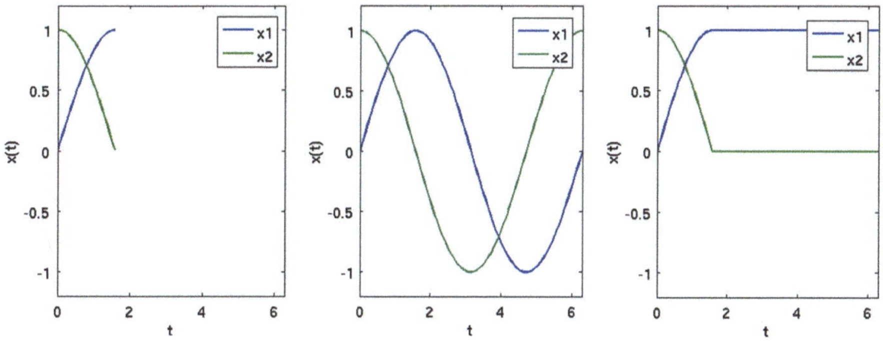

Fig. 9 Solutions of (27) for $x_1(0) = 0$ with Matlab, *OpenModelica* and *Dymola*

(iii) The most curious behavior was obtained with, e.g., Dymola using the integrators DOPRI45 – order 5 and Radau IIa – order 5 stiff and the default tolerance (1e-4). For $t \leq \frac{\pi}{2}$ it corresponds to the trigonometric functions, for $t \geq \frac{\pi}{2}$ to the constant solution $x_1 = 1, x_2 = 0$. In Matlab, we also obtain this behavior with ode23t for RelTol=AbsTol=1e-4.

However, none of these integrators indicated that there is a bifurcation at $t = \frac{\pi}{2}$. In contrast, *daePAD* detects and reports that a singularity occurs, since G_1 from the matrix sequence becomes singular.

Let us finally consider an example from applications.

Example 7 (Catalyst mixing) The DAE is the necessary condition for optimality for an optimal control problem from chemical kinetics. More background is provided in [5]. The development of the DAE is given in [3], where more general optimal control problems and their resulting DAEs are considered.

For the variable $p = (p_1, p_2)$ the notation $p^+ = \max(0, p)$, $p^- = \max(-p, 0)$ is used. For $x = (z^a, z^b, v^a, v^b, F\ p_1, p_2)^T$ the semi explicit nonlinear DAE reads

$$\begin{aligned}
\frac{dz^a}{dt} &= (10z^b - z^a)F, \\
\frac{dz^b}{dt} &= (z^a - 10z^b)F - (1 - F)z_b, \\
\frac{dv^a}{dt} &= (v^a - v^b)F, \\
\frac{dv^b}{dt} &= 10(v^b - v^a)F + (v^b + 1)(1 - F), \\
0 &= (z^a - 10z^b)(v^a - v^b) - z^b(v^b + 1) + p_1^- - p_2^-, \\
0 &= p_1^+ - F, \\
0 &= p_2^+ - 1 + F.
\end{aligned} \tag{28}$$

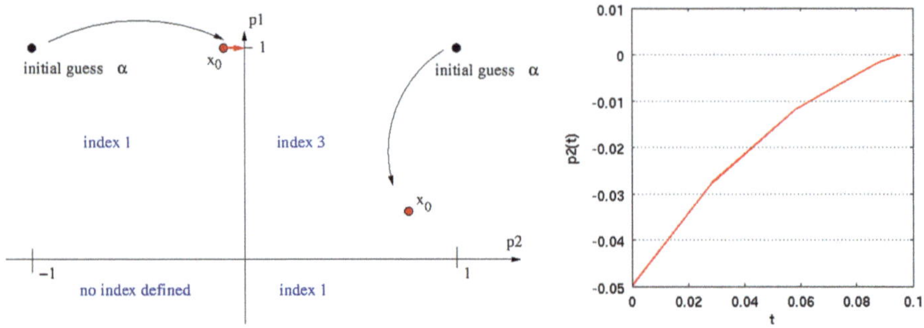

Fig. 10 Regularity regions of Example 7 (*left*) and component p_2 inside the index-1 region (*right*)

The analytical solution is known, which predestines this example for comparisons. The index of Example 7 was investigated in [2] and it was shown that (28) has index 1, if p_1 and p_2 are not zero and have different signs, and index 3 if both are greater than zero. For p_1 and p_2 smaller than zero no index is defined, because we observe a contradiction in the last two equations of (28) (cf. Fig. 10).

For $\alpha = \left(1, 0, -0.02, -0.07, 1, 1, -1\right)^T$, we start inside of one of the index-1 regions. In this case, the index is correctly determined, consistent initial values are computed and the integration is successful up to $t = 0.095598$. At this point, we reach with $p_2 = -3.22\text{e-}12$ the border of the regularity region and the integration stops with "minimal stepsize reached" (cf. Fig. 10). The border of the regularity region is here discovered by the index change (from 1 to 3), i.e., a rank change of G_1, if the stepsize is too large.

If we start inside the index-3 region with $\alpha = \left(1, 0, -0.02, -0.07, 1, 1, 1\right)$, the index is correctly determined, consistent initial values are computed and the integrations is successful. The computed constant singular control

$$F = 0.22714208270849801$$

is exactly the value of the theoretical result $\frac{9}{2(4+5\sqrt{10})}$.

We want to emphasize that the computation of consistent initial values for the index-3 case and the integration of the DAE in its index-3 form is a big step forward. Nevertheless our computational costs are not competitive for structured index 1 or 2 DAEs, since no structural assumptions are exploited. The computational times of several examples are reported in [4].

8 Outlook

The unique solvability of initial value problems of ODEs and DAEs is one of the basic properties we require for solving such problems numerically. Especially we need uniqueness for simulations from applications in order to compare the approximations with measurements. At a singularity, one of the properties like the existence, the boundedness or the uniqueness of a solution is always violated. Therefore, the detection of singularities of DAEs is of crucial importance for computations of results relevant in practice.

In our opinion, insufficient attention has been paid to the detection of singularities of ODEs/DAEs so far. If multiple solutions intersect at a point, then numerical integration packages usually follow one arbitrary solution, without any kind of warning. Mathematically, essential theoretical assumptions are not given at such points.

Due to its accuracy, AD opens new possibilities to diagnose such critical points while integrating ODEs/DAEs. In fact, the presented approach allows a verification of characteristic values related to the projector based analysis of DAEs, facilitating the detection of singularities. Particularly, it allows a point-wise index determination, a computation of consistent initial values and the integration of DAEs of index up to 3 monitoring singularities.

Acknowledgements The authors are indebted to Roswitha März for many fruitful discussions.

References

1. Dokchan, R.: Numerical intergration of differential-algebraic equations with harmless critical points. PhD thesis, Humboldt-University of Berlin (2011)
2. England, R., Gómez, S., Lamour, R.: Expressing optimal control problems as differential algebraic equations. Comput. Chem. Eng. **29**(8), 1720–1730 (2005)
3. England, R., Gómez, S., Lamour, R.: The properties of differential-algebraic equations representing optimal control problems. Appl. Numer. Math. **59**(10), 2357–2373 (2009)
4. Estévez Schwarz, D., Lamour, R.: Projector based integration of DAEs with the Taylor series method using automatic differentiation. J. Comput. Appl. Math. **262**, 62–72 (2014)
5. Jackson, R.: Optimal use of mixed catalyst for two successive chemical reactions. J. Optim. Theory Appl. **2**(1), 27–39 (1968)
6. Lamour, R., März, R., Tischendorf, C.: Differential-Algebraic Equations: A Projector Based Analysis. Differential-Algebraic Equations Forum 1. Springer, Berlin (2013)
7. März, R., Riaza, R.: Linear index-1 DAEs: regular and singular problems. Acta Appl. Math. **84**, 29–53 (2004)
8. März, R., Riaza, R.: Linear differential-algebraic equations with properly stated leading term: a-critical points. Math. Comput. Model. Dyn. Syst. **13**(3), 291–314 (2007)
9. Reissig, G., Martinson, W.S., Barton, P.I.: Differential-algebraic equations of index 1 may have an arbitrarily high structural index. SIAM J. Sci. Comput. **21**(6), 1987–1990 (2000)
10. Riaza, R.: Differential-Algebraic Systems. Analytical Aspects and Circuit Applications. World Scientific, Hackensack (2008)
11. Rump, S.M.: INTLAB – INTerval LABoratory. In: Csendes, T. (ed.) Developments in Reliable Computing, SCAN-98, pp. 77–104. Kluwer Academic Publishers, Dordrecht (1999)

Modular Time Integration of Block-Structured Coupled Systems Without Algebraic Loops

Martin Arnold

Abstract In March 2013, a Workshop on coupled Descriptor Systems was held at Castle Eringerfeld, Geseke (Germany). There, the author of this short communication presented a key note lecture on "Efficient time integration of block-structured descriptor systems" with results that have recently been published in Arnold et al. (Arch Mech Eng LX:75–94 2013) and are given also in Chap. 6 below. In the present short communication we give a compact introduction to this material and focus on basic aspects of the convergence analysis. For a more detailed discussion the interested reader is referred to Chap.6 or the abovementioned paper.

Keywords Co-simulation • Modular time integration • Algebraic loops • Convergence analysis

Mathematics Subject Classification (2010) 65L80 · 65L70

1 Co-simulation and Modular Time Integration

In engineering, co-simulation is a well established simulation technique for coupled time dependent problems. In nonlinear system dynamics, the basic theoretical and practical aspects of co-simulation found much interest in the late 1990s, see, e.g., the survey paper of Veitl et al. [19].

From the mathematical viewpoint, co-simulation results in a special class of modular time integration methods for initial value problems on finite time intervals $[t_0, t_{\text{end}}]$ that restrict the data exchange between subsystems to a finite number of communication points T_n, $(0 \leq n \leq N)$, with $t_0 = T_0 < T_1 < \ldots < T_N = t_{\text{end}}$,

M. Arnold (✉)
Institute of Mathematics, Martin Luther University Halle-Wittenberg, D-06099 Halle (Saale), Germany
e-mail: martin.arnold@mathematik.uni-halle.de

S. Schöps et al. (eds.), *Progress in Differential-Algebraic Equations*,
Differential-Algebraic Equations Forum, DOI 10.1007/978-3-662-44926-4_5

see [3]. Modular time integration proceeds in communication steps $T_n \to T_{n+1} = T_n + H_n$ of communication step size $H_n > 0$ from the initial state at $T_0 = t_0$ to the end time $t_{\text{end}} = T_N$.

In each communication step, the subsystems of the coupled problem are solved separately with different methods and different (micro) step sizes that may be tailored to the individual subsystems. The restriction of data exchange to discrete communication points T_n implies that *within* each communication step, i.e., for $t \in (T_n, T_{n+1})$, all terms of the coupled problem that represent the coupling of subsystems have to be approximated by appropriate extrapolation or interpolation techniques.

This data extrapolation causes a generic risk of exponential instability in the modular time integration of coupled differential-algebraic equations (DAE's), see [5, 15] and the more recent results in [2, 6, 16]. For communication step sizes $H_n \in (0, \bar{H}]$ with a sufficiently small constant $\bar{H} > 0$, this strange convergence behaviour in the DAE case has no equivalent in the modular time integration of coupled ordinary differential equations (ODE's) since extrapolation and interpolation of coupling terms introduce small perturbations in the right hand sides and the solution of ODE initial value problems with Lipschitz continuous right hand side depends continuously on such perturbations [20], see also the convergence result for modular time integration methods in [1].

From the theoretical viewpoint, the exponential instability of DAE modular time integration may always be avoided by linear or nonlinear stabilization methods that have been developed more than 10 years ago [5, 15, 18]. But in industrial applications problem specific stabilization techniques have been preferred instead or, alternatively, the model equations are re-formulated to avoid any constraint coupling in co-simulation.

Recently, the development of the interface standard Functional Mock-Up Interface (FMI) for Model Exchange and Co-Simulation [13] inspired new research on theoretical aspects of co-simulation. These theoretical investigations contribute to the ultimate goal that co-simulation techniques should achieve the level of robustness and numerical efficiency that is typical of classical monolithic simulation techniques. In that context, we studied the optimal choice of communication step sizes H_n and appropriate step size control algorithms [17].

2 Block-Structured Coupled Systems

In nonlinear system dynamics, communication step size control goes back to the work of Kübler [14] who used local error estimates based on Richardson extrapolation. This approach assumes implicitly that classical results from ODE and DAE theory apply as well to co-simulation and to modular time integration: (a) On finite time intervals, the global error is bounded in terms of the local errors (if an appropriate Lipschitz condition is satisfied) and (b) Richardson extrapolation gives an asymptotically correct estimate of the local error.

New results on order reduction in modular time integration [8] show that it is non-trivial to verify Assumption (a) in the block-oriented framework for co-simulation according to [14, 15]. In this block-oriented framework, the force-displacement coupling of two oscillators results in order reduction of the local error, see [8, 9]. There is no order reduction of the local error if the two oscillators are connected by displacement-displacement coupling which is, however, a less common approach in practical applications.

Error propagation and error accumulation in the modular time integration of block-structured coupled systems may be studied by standard techniques of the convergence analysis for one-step and multi-step methods in ODE and DAE time integration [4], see also [15]: Let the coupled system be composed of $r \geq 2$ subsystems

$$\left.\begin{aligned} \dot{\boldsymbol{x}}_j(t) &= \boldsymbol{f}_j(\boldsymbol{x}_j(t), \boldsymbol{u}_j(t), \boldsymbol{u}_{\text{ex}}(t)) \\ \boldsymbol{y}_j(t) &= \boldsymbol{g}_j(\boldsymbol{x}_j(t), \boldsymbol{u}_j(t)) \end{aligned}\right\} \tag{1a}$$

with $\boldsymbol{x}_j$, $\boldsymbol{u}_j$, $\boldsymbol{y}_j$, $(j = 1, \ldots, r)$, denoting the state, input and output vectors and with external inputs $\boldsymbol{u}_{\text{ex}}(t)$. The subsystems (1a) are coupled by input-output relations

$$\boldsymbol{u}_j(t) = \boldsymbol{c}_j\big(\boldsymbol{y}_1(t), \ldots, \boldsymbol{y}_{j-1}(t), \boldsymbol{y}_{j+1}(t), \ldots, \boldsymbol{y}_r(t)\big), \quad (j = 1, \ldots, r). \tag{1b}$$

Following line by line the analysis in [4], we summarize all components $\boldsymbol{x}_j$, $\boldsymbol{y}_j$, $\boldsymbol{u}_j$, $\boldsymbol{f}_j$, $\boldsymbol{g}_j$, $\boldsymbol{c}_j$ in vector form and get the coupled system

$$\left.\begin{aligned} \dot{\boldsymbol{x}}(t) &= \boldsymbol{f}(\boldsymbol{x}(t), \boldsymbol{u}(t), \boldsymbol{u}_{\text{ex}}(t)) \\ \boldsymbol{y}(t) &= \boldsymbol{g}(\boldsymbol{x}(t), \boldsymbol{u}(t)), \quad \boldsymbol{u}(t) = \boldsymbol{c}(\boldsymbol{y}(t)) \end{aligned}\right\} \tag{2}$$

with $\boldsymbol{x}(t) := \big(\boldsymbol{x}_1^{\top}(t), \boldsymbol{x}_2^{\top}(t), \ldots, \boldsymbol{x}_r^{\top}(t)\big)^{\top}, \ldots$. Equations (2) form a differential-algebraic equation (DAE) in variables $\boldsymbol{x}$, $\boldsymbol{y}$ and $\boldsymbol{u}$ that may, however, be reduced straightforwardly to an ordinary differential equation

$$\dot{\boldsymbol{x}}(t) = \boldsymbol{F}\big(\boldsymbol{x}(t), \boldsymbol{u}_{\text{ex}}(t)\big) := \boldsymbol{f}\big(\boldsymbol{x}(t), \boldsymbol{c}(\boldsymbol{g}(\boldsymbol{x}(t))), \boldsymbol{u}_{\text{ex}}(t)\big) \tag{3a}$$

and separate output equations

$$\boldsymbol{y}(t) = \boldsymbol{g}(\boldsymbol{x}(t)), \quad \boldsymbol{u}(t) = \boldsymbol{c}\big(\boldsymbol{g}(\boldsymbol{x}(t))\big) \tag{3b}$$

whenever there is no *direct feed-through* in the subsystems [15], i.e., whenever

$$\boldsymbol{y}_j(t) = \boldsymbol{g}_j(\boldsymbol{x}_j(t)), \quad \frac{\partial \boldsymbol{g}_j}{\partial \boldsymbol{u}_j}(\boldsymbol{x}_j, \boldsymbol{u}_j) \equiv \boldsymbol{0}, \ (j = 1, \ldots, r). \tag{4}$$

Modular time integration of (1) results in a numerical solution $(\tilde{\boldsymbol{x}}_j(t), \tilde{\boldsymbol{y}}_j(t), \tilde{\boldsymbol{u}}_j(t))$, $(j = 1, \dots, r)$, that is initialized by $\tilde{\boldsymbol{x}}_j(t_0) = \boldsymbol{x}_j(t_0)$ and solves in each communication step $T_n \to T_{n+1} = T_n + H_n$ a local initial value problem

$$\left.\begin{aligned} \dot{\tilde{\boldsymbol{x}}}_j(t) &= \boldsymbol{f}_j(\tilde{\boldsymbol{x}}_j(t), \tilde{\boldsymbol{u}}_j(t), \boldsymbol{u}_{\text{ex}}(t)) \\ \tilde{\boldsymbol{y}}_j(t) &= \boldsymbol{g}_j(\tilde{\boldsymbol{x}}_j(t), \tilde{\boldsymbol{u}}_j(t)) \end{aligned}\right\} \tag{5a}$$

with

$$\tilde{\boldsymbol{u}}_j(T_m) = \boldsymbol{c}_j\big(\tilde{\boldsymbol{y}}_1(T_m), \dots, \tilde{\boldsymbol{y}}_{j-1}(T_m), \tilde{\boldsymbol{y}}_{j+1}(T_m), \dots, \tilde{\boldsymbol{y}}_r(T_m)\big), \tag{5b}$$

$(j = 1, \dots, r;\ 0 \le m \le N)$, and approximations $\tilde{\boldsymbol{u}}_j(t) \approx \boldsymbol{u}(t)$, $(t \in (T_n, T_{n+1}))$, that are obtained by extrapolation of previously computed data $\{\tilde{\boldsymbol{u}}_j(T_m) : 0 \le m \le n\}$ or by interpolation of $\{\tilde{\boldsymbol{u}}_j(T_m) : 0 \le m \le n+1\}$ if $\tilde{\boldsymbol{u}}_j(T_m)$ with $m = n+1$ may be evaluated by (5b) before solving (5a) on $[T_n, T_{n+1}]$. Following the classical convergence analysis in [5], we focus on the error terms that result from data extrapolation and data interpolation in modular time integration and abstract from discretization errors in the numerical solution of the local initial value problems (5a).

Data interpolation is a frequently used approach in Gauss-Seidel like modular time integration methods for $r = 2$ coupled subsystems [5–7, 16] since in this special case the coupling condition (1b) for $j = 2$ simplifies to $\boldsymbol{u}_2(t) = \boldsymbol{c}_2(\boldsymbol{y}_1(t))$ and $\tilde{\boldsymbol{u}}_2(T_{n+1})$ may already be evaluated after solving (5a) on $[T_n, T_{n+1}]$ for the first subsystem $(j = 1)$. In the convergence analysis, we restrict ourselves, however, to Jacobi like methods with data extrapolation based on interpolation polynomials to keep the notation compact. Similar to the approach in [6], the results may be generalized to more complex co-simulation techniques including dynamic iteration schemes.

3 Local Error Analysis

For local error analysis, we consider functions $(\hat{\boldsymbol{x}}_j(t), \hat{\boldsymbol{y}}_j(t))$ solving (5a) on $[T_n, T_{n+1}]$ with *exact* initial values $\hat{\boldsymbol{x}}_j(T_n) = \boldsymbol{x}_j(T_n)$ and a polynomial subsystem input $\tilde{\boldsymbol{u}}_j(t)$ of degree $\le k$ that interpolates the *exact* subsystem input $\boldsymbol{u}_j(t)$ at the $k+1$ previous communication points $T_{n-\iota}$, $(\iota = 0, 1, \dots, k)$. In the convergence analysis, the local and global errors are studied for fixed communication step sizes $H_n = H$, i.e., $T_{n-\iota} = T_n - \iota H$ and $T_{n+1} = T_n + H$.

Lemma 1 *For coupled problems (2) with sufficiently often differentiable functions* $\boldsymbol{f}, \boldsymbol{g}, \boldsymbol{c}$, *polynomial data extrapolation with* $\tilde{\boldsymbol{u}}_j(T_{n-\iota}) = \boldsymbol{u}(T_{n-\iota})$, $(\iota = 0, 1, \dots, k)$, *results in* $\tilde{\boldsymbol{u}}_j(t) = \boldsymbol{u}_j(t) + \mathcal{O}(H^{k+1})$, $(t \in [T_n, T_{n+1}))$, *and local errors*

$$\|\hat{\boldsymbol{x}}_j(t) - \boldsymbol{x}_j(t)\| = \mathcal{O}(H^{k+2}), \quad \|\hat{\boldsymbol{y}}_j(t) - \boldsymbol{y}_j(t)\| = \mathcal{O}(H^{k+1}), \quad (t \in [T_n, T_{n+1}]). \tag{6}$$

For subsystems without direct feed-through, see (4), the error bound for the output vector $\mathbf{y}_j$ *may be improved by a factor of* H*:*

$$\|\hat{\mathbf{y}}_j(t) - \mathbf{y}_j(t)\| = \mathcal{O}(H^{k+2}), \; (t \in [T_n, T_{n+1}]). \tag{7}$$

Proof For any $t \in [T_n, T_n + H)$, there is a $\tau \in [T_n - kH, T_n + H]$, such that the interpolation error is given by

$$\tilde{\boldsymbol{u}}_j(t) - \boldsymbol{u}_j(t) = \frac{1}{(k+1)!} \frac{\mathrm{d}^{k+1}\boldsymbol{u}_j(\tau)}{\mathrm{d}t^{k+1}} \cdot \prod_{\iota=0}^{k} (t - (T_n - \iota H)) = \mathcal{O}(H^{k+1}), \tag{8}$$

see [12]. Since functions $\boldsymbol{x}_j(t)$ and $\hat{\boldsymbol{x}}_j(t)$ solve on $[T_n, T_n + H]$ initial value problems (1a), (5a) with one and the same initial value $\boldsymbol{x}_j(T_n)$, their difference is bounded by

$$\|\hat{\boldsymbol{x}}_j(t) - \boldsymbol{x}_j(t)\| \le C(\mathrm{e}^{L_j(t-T_n)} - 1) \max_{s \in [T_n, T_n+H)} \|\tilde{\boldsymbol{u}}_j(s) - \boldsymbol{u}_j(s)\|$$

with the Lipschitz constant L_j of $\boldsymbol{f}_j$ w.r.t. $\boldsymbol{x}_j$ and an appropriate constant $C > 0$, see, e.g., [20]. For $t \in [T_n, T_n + H]$, this estimate proves (6) for the state vector $\hat{\boldsymbol{x}}_j(t)$ and implies the corresponding error estimate for the output vector $\hat{\boldsymbol{y}}_j(t)$ that is defined in terms of $\hat{\boldsymbol{x}}_j(t)$ and $\tilde{\boldsymbol{u}}_j(t)$, see (1a) and (5a). To verify (7), we observe that condition (4) implies that the error in the output vector $\hat{\boldsymbol{y}}_j(t)$ is of the same size as the one in the state vector $\hat{\boldsymbol{x}}_j(t)$. □

Lemma 1 shows that constant/linear/quadratic polynomial interpolation of subsystem inputs results for systems without direct feed-through in local errors of size $\mathcal{O}(H^2)/\mathcal{O}(H^3)/\mathcal{O}(H^4)$ in the output vector $\boldsymbol{y}$. For systems with direct feed-through, an order reduction is observed and the local errors are only of order $\mathcal{O}(H)/\mathcal{O}(H^2)/\mathcal{O}(H^3)$. These theoretical results are in perfect agreement with the numerical test results of Busch [8] for two oscillators that are connected by displacement-displacement coupling (no direct feed-through) or by force-displacement coupling (direct feed-through).

4 Convergence Analysis

For the global error analysis, the output equations in (2) are formally eliminated resulting in a system of n_u (coupled) nonlinear equations

$$\boldsymbol{u} = \boldsymbol{c}\big(\boldsymbol{g}(\boldsymbol{x}, \boldsymbol{u})\big) \tag{9}$$

in the n_u (scalar) components u^ν, $(\nu = 1, \dots, n_u)$, of

$$\boldsymbol{u} = (\boldsymbol{u}_1^\top, \dots, \boldsymbol{u}_r^\top)^\top = (u^1, \dots, u^{n_u})^\top \in \mathbb{R}^{n_u}.$$

The coupled error propagation in the global errors $\boldsymbol{\varepsilon}^{x}(t) := \tilde{\boldsymbol{x}}(t) - \boldsymbol{x}(t)$ and $\boldsymbol{\varepsilon}^{u}(t) := \tilde{\boldsymbol{u}}(t) - \boldsymbol{u}(t)$ is studied at the communication points $t = T_n$:

Lemma 2 *With the assumptions of Lemma 1, the global errors satisfy*

$$\boldsymbol{\varepsilon}^{x}(T_{n+1}) = \boldsymbol{\varepsilon}^{x}(T_n) + \mathscr{O}(H)\Big(\|\boldsymbol{\varepsilon}^{x}(T_n)\| + \sum_{\iota=0}^{k} \|\boldsymbol{\varepsilon}^{u}(T_{n-\iota})\| \Big) + \mathscr{O}(H^{k+2}), \tag{10a}$$

$$\boldsymbol{\varepsilon}^{u}(T_{n+1}) = \frac{\partial \boldsymbol{c}\big(\boldsymbol{g}(\boldsymbol{x},\boldsymbol{u})\big)}{\partial \boldsymbol{u}}\Big|_{\boldsymbol{x}=\boldsymbol{x}(T_n),\,\boldsymbol{u}=\boldsymbol{u}(T_n)} \sum_{\iota=0}^{k} L_{\iota}^{(n,k)}(T_{n+1})\, \boldsymbol{\varepsilon}^{u}(T_{n-\iota}) + \tag{10b}$$

$$+ \mathscr{O}(1)\|\boldsymbol{\varepsilon}^{x}(T_{n+1})\| + \mathscr{O}(H)\sum_{\iota=0}^{k} \|\boldsymbol{\varepsilon}^{u}(T_{n-\iota})\| + \mathscr{O}(H^{k+1}) \tag{10c}$$

with Lagrange polynomials

$$L_{\iota}^{(n,k)}(t) := \prod_{\substack{l=0\\ l\neq\iota}}^{k} \frac{t - T_{n-l}}{T_{n-\iota} - T_{n-l}} = \prod_{\substack{l=0\\ l\neq\iota}}^{k} \frac{l + (t - T_n)/H}{l - \iota}\,.$$

Proof Using the Lagrange representation of interpolation polynomials [12], we get

$$\tilde{\boldsymbol{u}}(t) = \sum_{\iota=0}^{k} L_{\iota}^{(n,k)}(t)\, \tilde{\boldsymbol{u}}(T_{n-\iota})\,,\ \ (t \in [T_n, T_{n+1})\,)$$

and observe that the (formal) interpolation of the exact subsystem inputs $\boldsymbol{u}(t)$ results in

$$\boldsymbol{u}(t) = \sum_{\iota=0}^{k} L_{\iota}^{(n,k)}(t)\, \boldsymbol{u}(T_{n-\iota}) + \mathscr{O}(H^{k+1})\,,\ \ (t \in [T_n, T_{n+1})\,)$$

since the interpolation error is of size $\mathscr{O}(H^{k+1})$, see (8). Therefore, estimate (10b) is a straightforward consequence of (9):

$$\tilde{\boldsymbol{u}}(T_{n+1}) = \boldsymbol{c}\Big(\boldsymbol{g}\big(\tilde{\boldsymbol{x}}(T_{n+1}), \sum_{\iota=0}^{k} L_{\iota}^{(n,k)}(T_{n+1})\, \tilde{\boldsymbol{u}}(T_{n-\iota})\big)\Big)\,,$$

$$\boldsymbol{u}(T_{n+1}) = \boldsymbol{c}\Big(\boldsymbol{g}\big(\boldsymbol{x}(T_{n+1}), \sum_{\iota=0}^{k} L_{\iota}^{(n,k)}(T_{n+1})\, \boldsymbol{u}(T_{n-\iota}) + \mathscr{O}(H^{k+1})\big)\Big)\,.$$

To prove (10a), we consider

$$\begin{aligned}\boldsymbol{\varepsilon}^{x}(T_{n+1}) &= \tilde{\boldsymbol{x}}(T_{n+1}) - \hat{\boldsymbol{x}}(T_{n+1}) + \hat{\boldsymbol{x}}(T_{n+1}) - \boldsymbol{x}(T_{n+1}) \\ &= \tilde{\boldsymbol{x}}(T_{n+1}) - \hat{\boldsymbol{x}}(T_{n+1}) + \mathcal{O}(H^{k+1}),\end{aligned}$$

see (6). Functions $\tilde{\boldsymbol{x}}_j(t)$ and $\hat{\boldsymbol{x}}_j(t)$ solve at $[T_n, T_{n+1}]$ initial value problems (5a) with initial values $\tilde{\boldsymbol{x}}_j(T_n)$ and $\boldsymbol{x}_j(T_n)$ and right hand sides that differ by the definition of $\tilde{\boldsymbol{u}}_j(t)$: For $\tilde{\boldsymbol{x}}_j(t)$, we have $\tilde{\boldsymbol{u}}_j(t) = \sum_\iota L_\iota^{(n,k)}(t)\tilde{\boldsymbol{u}}_j(T_{n-\iota})$ and for $\hat{\boldsymbol{x}}_j(t)$ the approximated inputs are given by $\tilde{\boldsymbol{u}}_j(t) = \sum_\iota L_\iota^{(n,k)}(t)\boldsymbol{u}_j(T_{n-\iota})$, see Lemma 1. Therefore, $\tilde{\boldsymbol{x}}_j(T_{n+1}) - \hat{\boldsymbol{x}}_j(T_{n+1})$ depends continuously on $\tilde{\boldsymbol{x}}(T_n) - \boldsymbol{x}(T_n) = \boldsymbol{\varepsilon}^x(T_n)$ and on $\tilde{\boldsymbol{u}}(T_{n-\iota}) - \boldsymbol{u}(T_{n-\iota}) = \boldsymbol{\varepsilon}^u(T_{n-\iota})$, ($\iota = 0, 1, \ldots, k$), and estimate (10a) follows from classical results of the sensitivity analysis for ODE initial value problems [20]. □

It is well known that coupled error recursions like (10) may get unstable unless a contractivity condition is satisfied, see the convergence analysis in [11] and its application to modular time integration in [5]. As in [4], we restrict ourselves to coupled systems (1) being *free of algebraic loops* [15], i.e., we suppose that there is a permutation π of the index set $\{1, \ldots, n_{\boldsymbol{u}}\}$ such that for all $\nu = 1, \ldots, n_{\boldsymbol{u}}$ component u^{π_ν} of vector $\boldsymbol{u}$ in (9) is explicitly defined by the state vector $\boldsymbol{x}$ and by $\{u^{\pi_\mu} : 1 \le \mu < \nu\}$, i.e., $u^{\pi_\nu} = u^{\pi_\nu}(\boldsymbol{x}, u^{\pi_1}, \ldots, u^{\pi_{\nu-1}})$ does not depend on components u^{π_μ} with $\mu > \nu$. For a strict convergence analysis, we have to suppose that this permutation π may be defined independently of the current value of the state vector $\boldsymbol{x}$. This structural property is characterized by a directed graph that represents the structural nonzeros of the Jacobian $\partial\big(\boldsymbol{c}(\boldsymbol{g}(\boldsymbol{x},\boldsymbol{u}))\big)/\partial\boldsymbol{u}$ in (9):

Definition 1 The coupled system (1) is *free of algebraic loops in a structural sense* if there is an acyclic directed graph G with $n_{\boldsymbol{u}}$ vertices "ν" representing the $n_{\boldsymbol{u}}$ components u^ν of vector $\boldsymbol{u} \in \mathbb{R}^{n_{\boldsymbol{u}}}$ in (9) such that we have

$$\frac{\partial c^\nu\big(\boldsymbol{g}(\boldsymbol{x},\boldsymbol{u})\big)}{\partial u^\mu} \equiv 0$$

if there is no edge from vertex "ν" to vertex "μ" in G.

Definition 1 implies that the zero elements of the adjacency matrix $\mathbf{A}(G)$ represent structural zeros of the Jacobian $\partial\big(\boldsymbol{c}(\boldsymbol{g}(\boldsymbol{x},\boldsymbol{u}))\big)/\partial\boldsymbol{u}$ in (9). The matrix elements $a_{\nu\mu}$ of this adjacency matrix $\mathbf{A}(G) \in \{0,1\}^{n_{\boldsymbol{u}} \times n_{\boldsymbol{u}}}$ are nonzero iff there is an edge from vertex "ν" to vertex "μ" in G, see, e.g., [10, Appendix B.4] for a compact introduction to basics of graph theory. The adjacency matrix $\mathbf{A}(G)$ of an acyclic graph G is nilpotent and there is some $M \in \mathbb{N}$, $M \le n_u$, such that $(\mathbf{A}(G))^M = \mathbf{0}$. This fact implies for block-structured coupled systems (1) that are free of algebraic loops in a structural sense, see Definition 1, that

$$\prod_{l=1}^{M} \frac{\partial \boldsymbol{c}\big(\boldsymbol{g}(\boldsymbol{x},\boldsymbol{u})\big)}{\partial \boldsymbol{u}}\bigg|_{\boldsymbol{x}=\mathbf{x}^{[l]},\,\boldsymbol{u}=\mathbf{u}^{[l]}} \cdot \mathbf{Z}^{[l]} \equiv \mathbf{0} \tag{11}$$

for *any* system configurations $(\mathbf{x}^{[1]}, \mathbf{u}^{[1]}), \ldots, (\mathbf{x}^{[M]}, \mathbf{u}^{[M]})$ and any diagonal matrices $\mathbf{Z}^{[1]}, \ldots, \mathbf{Z}^{[M]} \in \mathbb{R}^{n_u \times n_u}$.

Iterating the error recursion (10b) for the algebraic solution components, we see that (11) guarantees that the contractivity condition of the convergence analysis according to Deuflhard, Hairer and Zugck [11] is trivially satisfied with contractivity constant $\alpha = 0$, see [4]. Therefore, the convergence result follows from classical arguments of the analysis of DAE time integration methods:

Theorem 1 *Consider a block-structured coupled system (1) that is free of algebraic loops in a structural sense, see Definition 1, and suppose that the assumptions of Lemma 1 are satisfied. There are positive constants* C *and* $\bar{H}$ *such that*

$$\|\boldsymbol{\varepsilon}^x(T_n)\| + \|\boldsymbol{\varepsilon}^u(T_n)\| \leq C\big(\max_{1\leq m\leq k} \|\tilde{\boldsymbol{x}}(T_m) - \boldsymbol{x}(T_m)\| + H^{k+1}\big) \tag{12}$$

for all $H \in (0, \bar{H}]$ *and for all* n *with* $t_0 + nH \leq t_{\text{end}}$ *provided that the starting values* $\big(\tilde{\boldsymbol{x}}(T_m), \tilde{\boldsymbol{u}}(T_m)\big)$, $(m = 0, 1, \ldots, k)$, *fulfill*

$$\|\tilde{\boldsymbol{x}}(T_m) - \boldsymbol{x}(T_m)\| = \mathcal{O}(H)\,, \quad \|\tilde{\boldsymbol{u}}(T_m) - \boldsymbol{c}\big(\boldsymbol{g}(\tilde{\boldsymbol{x}}(T_m), \tilde{\boldsymbol{u}}(T_m))\big)\| = \mathcal{O}(H^{k+1})\,.$$

For coupled systems without direct feed-through in the output equations, see (4), the assumptions of Theorem 1 are trivially satisfied with a graph G that has no edges at all and $\mathbf{A}(G) = \mathbf{0}$. The assumptions are satisfied as well in the case of subsystems being connected by a force-displacement coupling. Here, we observe direct feed-through in one of the output equations but the coupled system is nevertheless free of algebraic loops in the structural sense of Definition 1. For $M \geq 2$, the matrix products in (11) vanish identically [4].

As a practical consequence of this convergence theorem, we see that constant/linear/quadratic polynomial interpolation of subsystem inputs results in global errors of size $\mathcal{O}(H)/\mathcal{O}(H^2)/\mathcal{O}(H^3)$ whenever the coupled system is in a structural sense free of algebraic loops. The order reduction phenomenon that was observed for the local error in the case of force-displacement coupling [8] does not affect the order of the global error.

5 Summary

In the present short communication, we discussed the basic steps of a convergence analysis for the modular time integration of block-structured coupled systems that has recently been developed in [4]. We presented an easy-to-check sufficient criterion for the convergence of modular time integration methods that guarantees that on finite time intervals the global error may be bounded in terms of local errors. Furthermore, the global error does not suffer from order reduction phenomena that have recently been observed for the local error in the case of force-displacement coupling of mechanical systems [8]. The sufficient criterion makes use of methods from graph theory and may be interpreted as a condition on the coupled system to exclude algebraic loops in a structural sense. It is trivially satisfied for systems without direct feed-through in the output equations and may be verified also for mechanical systems being connected by a force-displacement coupling. For the latter problem class that is of special interest from the viewpoint of practical applications, no convergence results have been known before from the literature.

References

1. Arnold, M.: Multi-rate time integration for large scale multibody system models. In: Eberhard, P. (ed.) IUTAM Symposium on Multiscale Problems in Multibody System Contacts, Stuttgart, pp. 1–10. Springer (2007)
2. Arnold, M.: Stability of sequential modular time integration methods for coupled multibody system models. J. Comput. Nonlinear Dyn. **5**, 031003 (2010)
3. Arnold, M., Burgermeister, B., Führer, C., Hippmann, G., Rill, G.: Numerical methods in vehicle system dynamics: state of the art and current developments. Veh. Syst. Dyn. **49**, 1159–1207 (2011)
4. Arnold, M., Clauß, C., Schierz, T.: Error analysis and error estimates for co-simulation in FMI for Model Exchange and Co-Simulation v2.0. Arch. Mech. Eng. **LX**, 75–94 (2013)
5. Arnold, M., Günther, M.: Preconditioned dynamic iteration for coupled differential-algebraic systems. BIT Numer. Math. **41**, 1–25 (2001)
6. Bartel, A., Brunk, M., Günther, M., Schöps, S.: Dynamic iteration for coupled problems of electric circuits and distributed devices. SIAM J. Sci. Comput. **35**, B315–B335 (2013)
7. Burrage, K.: Parallel and Sequential Methods for Ordinary Differential Equations. Clarendon Press, Oxford (1995)
8. Busch, M.: Zur effizienten Kopplung von Simulationsprogrammen. PhD thesis, Universität Kassel, Fachbereich Maschinenbau (2012)
9. Busch, M., Schweizer, B.: Numerical stability and accuracy of different co-simulation techniques: analytical investigations based on a 2-DOF test model. In: Proceedings of the 1st Joint International Conference on Multibody System Dynamics, Lappeenranta, 25–27 May 2010
10. Cormen, T.H., Leiserson, C.E., Rivest, R.L., Stein, C.: Introduction to Algorithms, 2nd edn. MIT, Cambridge (2001)
11. Deuflhard, P., Hairer, E., Zugck, J.: One–step and extrapolation methods for differential–algebraic systems. Numer. Math. **51**, 501–516 (1987)
12. Deuflhard, P., Hohmann, A.: Numerical Analysis in Modern Scientific Computing: An Introduction, 2nd edn. Number 43 in Texts in Applied Mathematics. Springer, New York (2003)
13. FMI: The Functional Mockup Interface. https://www.fmi-standard.org/
14. Kübler, R.: Modulare Modellierung und Simulation mechatronischer Systeme. Fortschritt-Berichte VDI Reihe 20, Nr. 327. VDI–Verlag GmbH, Düsseldorf (2000)
15. Kübler, R, Schiehlen, W.: Two methods of simulator coupling. Math. Comput. Model. Dyn. Syst. **6**, 93–113 (2000)
16. Schierz, T., Arnold, M.: Stabilized overlapping modular time integration of coupled differential-algebraic equations. Appl. Numer. Math. **62**, 1491–1502 (2012)
17. Schierz, T., Arnold, M., Clauß, C.: Co-simulation with communication step size control in an FMI compatible master algorithm. In: Otter, M., Zimmer, D. (eds.) Proceedings of the 9th International Modelica Conference, Munich, 3–5 Sept 2012

18. Tseng, F.C., Hulbert, G.M.: A gluing algorithm for network-distributed multibody dynamics simulation. Multibody Syst. Dyn. **6**, 377–396 (2001)
19. Veitl, A., Gordon, T., van de Sand, A., Howell, M., Valášek, M., Vaculín, O., Steinbauer, P.: Methodologies for coupling simulation models and codes in mechatronic system analysis and design. In: Proceedings of the 16th IAVSD–Symposium on Dynamics of Vehicles on Roads and Tracks. Pretoria. Supplement to Vehicle System Dynamics, vol. 33, pp. 231–243. Swets & Zeitlinger (1999)
20. Walter, W.: Ordinary Differential Equations. Number 182 in Graduate Texts in Mathematics. Springer, New York (1998)

Error Analysis and Error Estimates for Co-simulation in FMI for Model Exchange and Co-Simulation v2.0

Martin Arnold, Christoph Clauß, and Tom Schierz

Abstract Complex multi-disciplinary models in system dynamics are typically composed of subsystems. This modular structure of the model reflects the modular structure of complex engineering systems. In industrial applications, the individual subsystems are often modelled separately in different mono-disciplinary simulation tools. The *Functional Mock-Up Interface* (FMI) provides an interface standard for coupling physical models from different domains and addresses problems like export and import of model components in industrial simulation tools (*FMI for Model Exchange*) and the standardization of co-simulation interfaces in nonlinear system dynamics (*FMI for Co-Simulation*), see https://www.fmi-standard.org/. The renewed interest in algorithmic and numerical aspects of co-simulation inspired some new investigations on error estimation and stabilization techniques in FMI for Model Exchange and Co-Simulation v2.0 compatible co-simulation environments. In the present paper, we focus on reliable error estimation for communication step size control in this framework.

Keywords Co-simulation • Modular time integration • Algebraic loops • Convergence analysis

---

This paper has been previously published in Archive of Mechanical Engineering LX(2013)75-94 (DOI: 10.2478/meceng-2013-0005) and is reproduced in the present proceedings volume with kind permission of the Polish Academy of Sciences.

M. Arnold (✉)
Institute of Mathematics, Martin Luther University Halle-Wittenberg, D-06099 Halle (Saale), Germany
e-mail: martin.arnold@mathematik.uni-halle.de

C. Clauß
Fraunhofer Institute for Integrated Circuits IIS, Design Automation Division EAS, Zeunerstr. 38, D-01069 Dresden, Germany
e-mail: christoph.clauss@eas.iis.fraunhofer.de

T. Schierz
SIMPACK AG, Friedrichshafener Strasse 1, D-82205 Gilching, Germany
e-mail: tom.schierz@simpack.de

S. Schöps et al. (eds.), *Progress in Differential-Algebraic Equations*, Differential-Algebraic Equations Forum, DOI 10.1007/978-3-662-44926-4_6

Mathematics Subject Classification (2010) 65L80 · 65L70

1 Introduction

Co-Simulation is a rather general approach to the simulation of coupled technical systems and coupled physical phenomena in engineering with focus on time-dependent problems. Theoretical and practical aspects of co-simulation were recently studied in the ITEA2 project MODELISAR (2008–2011). A *Functional Mock-Up Interface* (FMI) was developed that provides an interface standard for coupling physical models from different domains and addresses problems like export and import of model components in industrial simulation tools (*FMI for Model Exchange*) and the standardization of co-simulation interfaces in nonlinear system dynamics (*FMI for Co-Simulation*), see [10].

Today, the interface standard is maintained and further developed within the Modelica Association Project "FMI". In August 2012, the fourth β-version of FMI for Model Exchange and Co-Simulation v2.0 was released [15] that supports advanced numerical techniques in co-simulation. Well known industrial simulation tools for applied dynamics support Version 1.0 of this standard and plan to support the forthcoming Version 2.0 in the near future, see the "Tools" tab of website [10] for up-to-date information.

FMI for co-simulation is an interface standard for the solution of time-dependent coupled systems consisting of subsystems that are continuous in time (model components that are described by time-dependent differential equations) or time-discrete (e.g., discrete controllers). In a block representation of the coupled system, the time continuous subsystems are represented by blocks with (internal) state variables $\boldsymbol{x}_j(t)$ that are connected to other subsystems (blocks) of the coupled problem by subsystem inputs $\boldsymbol{u}_j(t)$ and subsystem outputs $\boldsymbol{y}_j(t)$, see Fig. 1 and Sect. 2 below. In this framework, the physical connections between subsystems are represented by mathematical coupling conditions between the inputs $\boldsymbol{u}(t)$ and the outputs $\boldsymbol{y}(t)$ of all subsystems [14, 15].

Co-Simulation exploits the modular structure of coupled problems in all stages of the simulation process with separate model setup, pre- and postprocessing for the individual subsystems in different simulation tools. During time integration, the simulation is again performed independently for all subsystems restricting the data exchange between subsystems to discrete communication points T_n, see [2]. In different contexts, the communication points T_n, the communication steps $T_n \to T_{n+1}$ and the communication step sizes $H_n := T_{n+1} - T_n$ are also known as sampling points (synchronization points), macro steps and sampling rates, respectively [15].

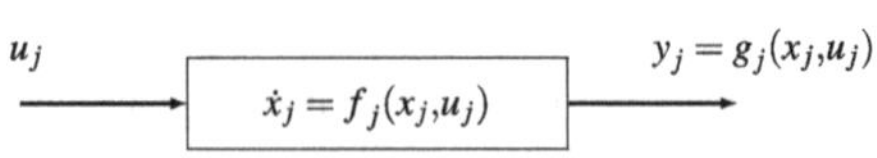

Fig. 1 Block-oriented representation of slave FMU's [15]

From the mathematical viewpoint, the coupling of different numerical solvers in a co-simulation environment results in a *modular* time integration method that substitutes the (unknown) subsystem inputs $\boldsymbol{u}_j(t)$ between the communication points T_n by some suitable approximation like polynomial extrapolation or interpolation [2]. This *signal extrapolation* introduces additional error terms in the modular time integration and may furthermore result in numerical instability. FMI for Model Exchange and Co-Simulation v2.0 addresses these problems by interface routines supporting higher order extrapolation and interpolation of subsystem inputs, communication step size control including step rejection and Jacobian based linearly implicit stabilization techniques.

The present paper focusses on communication step size control and error estimation. For more information about Jacobian based stabilization techniques we refer to [1] and the more recent results in [3, 17]. The paper is organized as follows: In Sect. 2, we give a sufficient criterion to exclude algebraic loops in a coupled system with block structure. Basic steps of a convergence analysis in this framework are presented in Sect. 3. Generalizing Richardson extrapolation techniques from ODE and DAE theory to modular time integration, we get reliable estimates for the local error. Theoretical and practical aspects are discussed in Sect. 4. The results of the theoretical analysis are verified by numerical tests for a quarter car model (Sect. 5). The paper ends with some conclusions in section "Conclusions".

The numerical tests for the quarter car benchmark problem were performed in a Matlab based test environment. Recently, these results were reproduced using an FMI for Co-Simulation v1.0 compatible master that was developed at Fraunhofer IIS/EAS within the MODELISAR project [7, 18].

This paper has been published before in *Archive of Mechanical Engineering* **LX**(2013)75–94 (DOI: 10.2478/meceng-2013-0005) and is reproduced in the present proceedings volume with kind permission of the Polish Academy of Sciences.

2 Block Representation of Coupled Systems

Following the approach of Kübler and Schiehlen [14], FMI for Model Exchange and Co-Simulation v2.0 is based on a block representation of coupled systems. The mathematical analysis of modular time integration methods for coupled systems of ordinary differential equations (ODEs) [1] may be extended to this block representation of coupled problems if there are no algebraic loops in the system. The specific numerical problems that may result from a coupling by constraints or other algebraic loops in the system found much interest in the early days of co-simulation algorithms in system dynamics, see, e.g., [4, 14], but have today only limited practical relevance in industrial applications.

Therefore, we will start the theoretical investigations by a structural analysis of coupled systems in block representation to exclude systems with algebraic loops. Let subsystem "j" be described by its state and output equations

$$\left.\begin{aligned} \dot{\boldsymbol{x}}_j(t) &= \boldsymbol{f}_j(\boldsymbol{x}_j(t), \boldsymbol{u}_j(t), \boldsymbol{u}_{\mathrm{ex}}(t)) \\ \boldsymbol{y}_j(t) &= \boldsymbol{g}_j(\boldsymbol{x}_j(t), \boldsymbol{u}_j(t)) \end{aligned}\right\} \tag{1a}$$

with $\boldsymbol{x}_j, \boldsymbol{u}_j, \boldsymbol{y}_j$ denoting the state, input and output vectors and some external input $\boldsymbol{u}_{\mathrm{ex}}(t)$. The $r \geq 2$ subsystems are coupled by input-output relations

$$\boldsymbol{u}_j(t) = \boldsymbol{c}_j\big(\boldsymbol{y}_1(t), \ldots, \boldsymbol{y}_{j-1}(t), \boldsymbol{y}_{j+1}(t), \ldots, \boldsymbol{y}_r(t)\big), \quad (j = 1, \ldots, r). \tag{1b}$$

Summarizing all components $\boldsymbol{x}_j$, $\boldsymbol{y}_j$, $\boldsymbol{u}_j$, $\boldsymbol{f}_j$, $\boldsymbol{g}_j$, $\boldsymbol{c}_j$ in vector form, we get the coupled system in the more compact form

$$\left.\begin{aligned} \dot{\boldsymbol{x}}(t) &= \boldsymbol{f}(\boldsymbol{x}(t), \boldsymbol{u}(t), \boldsymbol{u}_{\mathrm{ex}}(t)) \\ \boldsymbol{y}(t) &= \boldsymbol{g}(\boldsymbol{x}(t), \boldsymbol{u}(t)), \quad \boldsymbol{u}(t) = \boldsymbol{c}(\boldsymbol{y}(t)) \end{aligned}\right\} \tag{2}$$

with $\boldsymbol{x}(t) := \big(\boldsymbol{x}_1^\top(t), \boldsymbol{x}_2^\top(t), \ldots, \boldsymbol{x}_r^\top(t)\big)^\top, \ldots$. Equations (2) form a differential-algebraic equation (DAE) in variables $\boldsymbol{x}$, $\boldsymbol{y}$ and $\boldsymbol{u}$ that may, however, be reduced straightforwardly to an ordinary differential equation

$$\dot{\boldsymbol{x}}(t) = \boldsymbol{F}\big(\boldsymbol{x}(t), \boldsymbol{u}_{\mathrm{ex}}(t)\big) := \boldsymbol{f}\big(\boldsymbol{x}(t), \boldsymbol{c}(\boldsymbol{g}(\boldsymbol{x}(t))), \boldsymbol{u}_{\mathrm{ex}}(t)\big) \tag{3a}$$

and separate output equations

$$\boldsymbol{y}(t) = \boldsymbol{g}(\boldsymbol{x}(t)), \quad \boldsymbol{u}(t) = \boldsymbol{c}\big(\boldsymbol{g}(\boldsymbol{x}(t))\big) \tag{3b}$$

whenever there is no direct feed-through in the subsystems, i.e., whenever

$$\boldsymbol{y}_j(t) = \boldsymbol{g}_j(\boldsymbol{x}_j(t)), \quad \frac{\partial \boldsymbol{g}_j}{\partial \boldsymbol{u}_j}(\boldsymbol{x}_j, \boldsymbol{u}_j) \equiv \boldsymbol{0}, \; (j = 1, \ldots, r). \tag{4}$$

From the mathematical viewpoint, the ODE like structure (3) would be much more favourable than the general DAE system (2) but condition (4) is too restrictive and excludes, e.g., force-displacement couplings between two mechanical systems.

Example 1 (Busch and Schweizer [6]) Consider two 1-DOF oscillators with position and velocity coordinates (p_1, v_1) and (p_2, v_2), respectively, that are coupled by a linear spring with stiffness and damping parameters c and d resulting in a spring force

$$F_{\mathrm{c}} = c(p_2 - p_1) + d(v_2 - v_1).$$

Variables (p_j, v_j) are components of state vectors $\boldsymbol{x}_j$, $(j = 1, 2)$. A *displacement-displacement coupling* is realized by input-output couplings

$$\boldsymbol{u}_1 = (u_1^1, u_1^2)^\top = \boldsymbol{y}_2 \in \mathbb{R}^2\,,\quad \boldsymbol{u}_2 = (u_2^1, u_2^2)^\top = \boldsymbol{y}_1 \in \mathbb{R}^2$$

and subsystem outputs $\boldsymbol{y}_j = (p_j, v_j)^\top$, $(j = 1, 2)$, such that there is no direct feed-through and condition (4) is satisfied. In both subsystems "j", the coupling spring force is evaluated as $F_\mathrm{c} = (-1)^j \big(c(p_j - u_j^1) + d(v_j - u_j^2)\big)$.

The same physical system may also be described by a *force-displacement coupling* with input-output couplings

$$\boldsymbol{u}_1 = (u_1^1) = \boldsymbol{y}_2 \in \mathbb{R}\,,\quad \boldsymbol{u}_2 = (u_2^1, u_2^2)^\top = \boldsymbol{y}_1 \in \mathbb{R}^2$$

and subsystem outputs

$$\boldsymbol{y}_1 = \boldsymbol{g}_1(\boldsymbol{x}_1) = (p_1, v_1)^\top\,,\qquad \boldsymbol{y}_2 = \boldsymbol{g}_2(\boldsymbol{x}_2, \boldsymbol{u}_2) = F_\mathrm{c} = c(p_2 - u_2^1) + d(v_2 - u_2^2)$$

resulting in a direct feed-through in the second subsystem such that (4) is violated for $j = 2$. In the first subsystem, the coupling spring force is directly given by $\boldsymbol{u}_1 = \boldsymbol{y}_2 = F_\mathrm{c}$. In the second subsystem it is again evaluated as

$$F_\mathrm{c} = c(p_2 - u_2^1) + d(v_2 - u_2^2)\,.$$

In Example 1, the force-displacement coupling results in a slightly more complex coupling structure but the overall coupled system is (as before) free of algebraic loops. For coupled systems of the general form (2) with direct feed-through in one or more subsystems, a refined structural analysis is necessary to exclude algebraic loops. For this analysis, the output equations in (2) are formally eliminated resulting in a system of n_u (coupled) nonlinear equations

$$\boldsymbol{u} = \boldsymbol{c}\big(\boldsymbol{g}(\boldsymbol{x}, \boldsymbol{u})\big) \tag{5}$$

in the n_u (scalar) components u^ν, $(\nu = 1, \ldots, n_u)$, of

$$\boldsymbol{u} = (\boldsymbol{u}_1^\top, \ldots, \boldsymbol{u}_r^\top)^\top = (u^1, \ldots, u^{n_u})^\top \in \mathbb{R}^{n_u}\,.$$

In (5), component u^ν does not depend directly on component u^μ if $\partial c^\nu\big(\boldsymbol{g}(\boldsymbol{x}, \boldsymbol{u})\big)/\partial u^\mu \equiv 0$. In a structural sense, component u^ν may depend directly on component u^μ if

$$\left.\frac{\partial c^\nu\big(\boldsymbol{g}(\boldsymbol{x}, \boldsymbol{u})\big)}{\partial u^\mu}\right|_{\boldsymbol{x}=\boldsymbol{x}^*, \boldsymbol{u}=\boldsymbol{u}^*} \neq 0 \quad \text{for some argument } (\boldsymbol{x}^*, \boldsymbol{u}^*). \tag{6}$$

In that case, there is a *structural feed-through path of length* 1 from u^μ to u^ν. In Example 1, there are no structural feed-through paths in the case of displacement-displacement coupling. For force-displacement coupling, structural feed-through paths of length 1 point from u_2^1 to u_1^1 and from u_2^2 to u_1^1.

A *structural feed-through path of length* L from u^μ to u^ν consists of L connected structural feed-through paths of length 1. It is characterized by $L+1$ indices $\mu_l \in \{0, 1, \ldots, n_u\}$, $(l = 0, 1, \ldots, L)$, with $\mu_k \neq \mu_l$ whenever $0 < |k - l| < L$, $(k, l = 0, 1, \ldots, L)$, $\mu_0 = \mu$, $\mu_L = \nu$ and structural feed-through paths of length 1 from $u^{\mu_{l-1}}$ to u^{μ_l}, $(l = 1, \ldots, L)$. The structural feed-through path from u^μ to u^ν is called *open* if $\mu \neq \nu$ and *closed* otherwise. In (5), there are no closed structural feed-through paths of length $L = 1$ since subsystem input vector $\boldsymbol{u}_j$ does not depend directly on subsystem output vector $\boldsymbol{y}_j$, see (1b). A trivial closed structural feed-through path of length $L = 2$ would result from output equations $\boldsymbol{y}_1 = \boldsymbol{u}_1$, $\boldsymbol{y}_2 = \boldsymbol{u}_2$ with coupling conditions $\boldsymbol{u}_1 = \boldsymbol{y}_2$, $\boldsymbol{u}_2 = \boldsymbol{y}_1$ (algebraic loop). In Example 1, there are no structural feed-through paths of length $L \geq 2$ and therefore also no closed structural feed-through paths.

Structural feed-through paths in (5) may be studied conveniently by a directed graph G with n_u vertices "ν" that correspond to the components u^ν, $(\nu = 1, \ldots, n_u)$, of $\boldsymbol{u}$ in (5), see, e.g., [8, Appendix B.4] for a compact introduction to basics of graph theory. In this directed graph, an edge points from vertex "ν" to vertex "μ" whenever condition (6) is satisfied. Each structural feed-through path of length L in (5) corresponds to a path of length L in the associated graph G. On the other hand, a cycle in G represents a closed structural feed-through path in the system of nonlinear equations (5).

To exclude algebraic loops in the coupled system (2), we suppose in the following that the directed graph G associated with (5) is acyclic, i.e., free of cycles. This condition implies that the adjacency matrix $\mathbf{A}(G) \in \mathbb{R}^{n_u \times n_u}$ is nilpotent and there is some $M \in \mathbb{N}$, $M \leq n_u$, such that $(\mathbf{A}(G))^M = \mathbf{0}$, see [8, Chapter 22]. Here, the n_u^2 elements $a_{\nu\mu}(G)$ of $\mathbf{A}(G)$ are given by $a_{\nu\mu} = 1$ if there is an edge from vertex "ν" to vertex "μ" and $a_{\nu\mu} = 0$ otherwise. Condition (6) shows that the adjacency matrix represents the "structural" sparsity pattern of the Jacobian $\partial\big(\boldsymbol{c}(\boldsymbol{g}(\boldsymbol{x},\boldsymbol{u}))\big)/\partial\boldsymbol{u}$. As a practical consequence, the nilpotency of $\mathbf{A}(G)$ implies

$$\prod_{l=1}^{M} \frac{\partial \boldsymbol{c}\big(\boldsymbol{g}(\boldsymbol{x},\boldsymbol{u})\big)}{\partial \boldsymbol{u}}\bigg|_{\boldsymbol{x}=\mathbf{x}^{[l]},\,\boldsymbol{u}=\mathbf{u}^{[l]}} \cdot \mathbf{Z}^{[l]} \equiv \mathbf{0} \tag{7}$$

for *any* system configurations $(\mathbf{x}^{[1]}, \mathbf{u}^{[1]}), \ldots, (\mathbf{x}^{[M]}, \mathbf{u}^{[M]})$ and any diagonal matrices $\mathbf{Z}^{[1]}, \ldots, \mathbf{Z}^{[M]} \in \mathbb{R}^{n_u \times n_u}$.

Example 2 For systems without direct feed-through, graph G has no edges at all, see (4) and (6). Therefore, the displacement-displacement coupling in Example 1 results in the graph G in the left plot of Fig. 2 with $\mathbf{A}(G) = \mathbf{0}_{4\times4}$. For the force-displacement coupling, we obtain the graph G in the right plot of Fig. 2 and the non-zero but nilpotent adjacency matrix

$$\mathbf{A}(G) = \begin{pmatrix} 0 & 1 & 1 \\ 0 & 0 & 0 \\ 0 & 0 & 0 \end{pmatrix} \in \mathbb{R}^{3\times3}$$

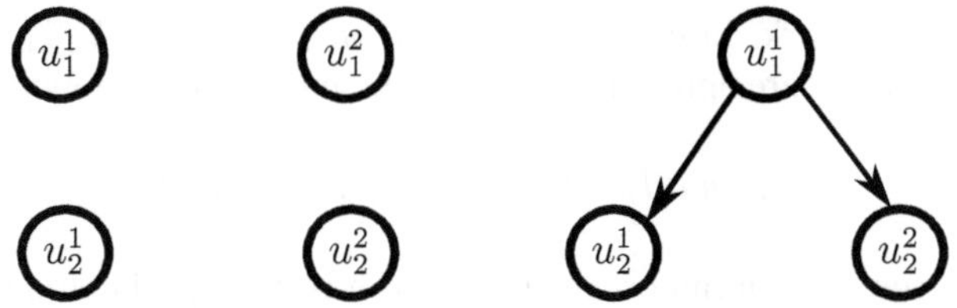

Fig. 2 Graphs representing the feed-through structure of the model in Example 1. *Left plot*: displacement-displacement coupling, *Right plot*: force-displacement coupling

with $(\mathbf{A}(G))^2 = \mathbf{0}$. Since output equations and coupling equations are linear, the Jacobian $\partial \boldsymbol{c}\big(\boldsymbol{g}(\boldsymbol{x},\boldsymbol{u})\big)/\partial \boldsymbol{u}$ is constant and we get

$$\frac{\partial \boldsymbol{c}\big(\boldsymbol{g}(\boldsymbol{x},\boldsymbol{u})\big)}{\partial \boldsymbol{u}}\mathbf{Z}^{[1]}\frac{\partial \boldsymbol{c}\big(\boldsymbol{g}(\boldsymbol{x},\boldsymbol{u})\big)}{\partial \boldsymbol{u}}\mathbf{Z}^{[2]} = \begin{pmatrix} 0 & -c & -d \\ 0 & 0 & 0 \\ 0 & 0 & 0 \end{pmatrix}\mathbf{Z}^{[1]}\begin{pmatrix} 0 & -c & -d \\ 0 & 0 & 0 \\ 0 & 0 & 0 \end{pmatrix}\mathbf{Z}^{[2]}$$

$$= \mathbf{0}_{3\times 3} \tag{8}$$

for any diagonal matrices $\mathbf{Z}^{[1]}$, $\mathbf{Z}^{[2]} \in \mathbb{R}^{3\times 3}$.

3 Convergence Analysis

Standard step size control algorithms in ODE and DAE time integration adjust the step size in each time step to guarantee that (an estimate) of the *local* error does not exceed some user defined error bounds. A perturbation analysis shows that this strategy will bound the *global* error after a finite time interval of length T as well since local errors are amplified in time integration by a factor that is bounded by $\exp(L_0 T)$ with L_0 denoting the Lipschitz constant of the right hand side (in the ODE case). For coupled systems in block representation (1), the situation is more complicated since order reduction phenomena have been reported for the local error analysis [5, Appendix 2.A] that do, however, not deteriorate the practically observed order of the global error.

In the present section, we discuss basic steps of a convergence analysis for modular time integration methods applied to coupled systems in block representation (1) that provides a theoretical justification of this practically observed error behaviour. Local and global errors are not as directly connected as in the classical ODE and DAE case but the communication step size control may nevertheless be based on local error estimates.

For the theoretical analysis, we follow the framework of [4] and neglect the discretization errors of the time integration in all subsystems to focus on the additional error terms that are introduced by coupling these subsystems in a co-simulation framework. In each communication step $T_n \to T_{n+1} = T_n + H$, the input vectors $\boldsymbol{u}_j(t)$ are approximated by some interpolation polynomial

$$\boldsymbol{\Psi}_j\big(t; \boldsymbol{u}_j(T_{n-k}), \ldots, \boldsymbol{u}_j(T_n), \boldsymbol{u}_j(T_{n+1})\big)$$

of degree $\leq k$, see [9]. An important special case is the signal extrapolation based on interpolation polynomials

$$\boldsymbol{\Psi}_j(t) = \sum_{\iota=0}^{k} \boldsymbol{u}_j(T_{n-\iota}) \prod_{\substack{l=0 \\ l \neq \iota}}^{k} \frac{t - T_{n-l}}{T_{n-\iota} - T_{n-l}} = \boldsymbol{u}_j(t) + \mathcal{O}(H^{k+1}) \,. \tag{9}$$

Example 3 In a brute force approach, the input vectors $\boldsymbol{u}_j(t)$ are simply frozen in each communication step $T_n \to T_{n+1} = T_n + H$ resulting in $k = 0$ and a constant extrapolation polynomial $\boldsymbol{\Psi}_j(t) \equiv \boldsymbol{u}_j(T_n)$, $(t \in [T_n, T_{n+1}])$. For higher accuracy, linear ($k = 1$) and quadratic ($k = 2$) interpolation is more favourable:

$$\boldsymbol{\Psi}_j(t) = \boldsymbol{u}_j(T_n) + \frac{\boldsymbol{u}_j(T_n) - \boldsymbol{u}_j(T_{n-1})}{T_n - T_{n-1}}(t - T_n) \,, \quad (k = 1) \,,$$

$$\boldsymbol{\Psi}_j(t) = \boldsymbol{u}_j(T_n) + \frac{\boldsymbol{u}_j(T_n) - \boldsymbol{u}_j(T_{n-1})}{T_n - T_{n-1}}(t - T_n) + \boldsymbol{\beta}_{j,n}(t - T_n)(t - T_{n-1}) \,, \quad (k = 2) \,,$$

with

$$\boldsymbol{\beta}_{j,n} := \Big(\frac{\boldsymbol{u}_j(T_n) - \boldsymbol{u}_j(T_{n-1})}{T_n - T_{n-1}} - \frac{\boldsymbol{u}_j(T_{n-1}) - \boldsymbol{u}_j(T_{n-2})}{T_{n-1} - T_{n-2}}\Big) / (T_n - T_{n-2}) \,.$$

The local error in communication step $T_n \to T_{n+1} = T_n + H$ compares for all $t \in [T_n, T_{n+1}]$ the solution of (2) with the solution of

$$\left.\begin{aligned} \dot{\hat{\boldsymbol{x}}}(t) &= \boldsymbol{f}(\hat{\boldsymbol{x}}(t), \boldsymbol{\Psi}(t), \boldsymbol{u}_{\text{ex}}(t)) \\ \hat{\boldsymbol{y}}(t) &= \boldsymbol{g}(\hat{\boldsymbol{x}}(t), \boldsymbol{\Psi}(t)) \\ \hat{\boldsymbol{u}}(t) &= \boldsymbol{c}(\hat{\boldsymbol{y}}(t)) \end{aligned}\right\} \,, \quad (t \in [T_n, T_{n+1}]) \,, \quad \hat{\boldsymbol{x}}(T_n) = \boldsymbol{x}(T_n) \,. \tag{10}$$

The perturbation analysis for ODE initial value problems [19] shows

$$\begin{aligned} \|\hat{\boldsymbol{x}}(T_{n+1}) - \boldsymbol{x}(T_{n+1})\| &\leq C_x\big(\exp(L_0(T_{n+1} - T_n)) - 1\big) \max_{t \in [T_n, T_{n+1}]} \|\boldsymbol{\Psi}(t) - \boldsymbol{u}(t)\| \\ &= \mathcal{O}(H \cdot H^{k+1}) = \mathcal{O}(H^{k+2}) \,, \end{aligned}$$

$$\|\hat{\boldsymbol{u}}(T_{n+1}) - \boldsymbol{u}(T_{n+1})\| \leq C_u \|\hat{\boldsymbol{y}}(T_{n+1}) - \boldsymbol{y}(T_{n+1})\| = \mathcal{O}(H^{k+1}) \,,$$

see (9). Following the convergence analysis for linear multistep methods in the DAE case [11], we get a coupled error recursion for the global errors $\boldsymbol{\epsilon}_n^x$, $\boldsymbol{\epsilon}_n^u$ in differential and algebraic solution components:

$$\|\boldsymbol{\epsilon}_{n+1}^{\boldsymbol{x}}\| \leq \big(1+\mathcal{O}(H)\big)\|\boldsymbol{\epsilon}_{n}^{\boldsymbol{x}}\| + \mathcal{O}(H)\sum_{\iota=0}^{k}\|\boldsymbol{\epsilon}_{n-\iota}^{\boldsymbol{u}}\| + \mathcal{O}(H^{k+2})\,, \tag{11a}$$

$$\boldsymbol{\epsilon}_{n+1}^{\boldsymbol{u}} = \sum_{\iota=0}^{k}\mathbf{J}_n\mathbf{Z}_{n-\iota}\boldsymbol{\epsilon}_{n-\iota}^{\boldsymbol{u}} + \mathcal{O}(1)\|\boldsymbol{\epsilon}_{n}^{\boldsymbol{x}}\| + \mathcal{O}(H)\sum_{\iota=0}^{k}\|\boldsymbol{\epsilon}_{n-\iota}^{\boldsymbol{u}}\| + \mathcal{O}(H^{k+1}) \tag{11b}$$

with

$$\mathbf{J}_n := \left.\frac{\partial \boldsymbol{c}(\boldsymbol{g}(\boldsymbol{x},\boldsymbol{u}))}{\partial \boldsymbol{u}}\right|_{\boldsymbol{x}=\boldsymbol{x}(T_n),\,\boldsymbol{u}=\boldsymbol{u}(T_n)}, \qquad \mathbf{Z}_{n-\iota} := \left.\frac{\partial \boldsymbol{\Psi}}{\partial \boldsymbol{u}_{n-\iota}}(t;\boldsymbol{u}_{n-k},\ldots,\boldsymbol{u}_n,\boldsymbol{u}_{n+1})\right|_{t=T_{n+1}}.$$

The linear interpolation polynomial $\boldsymbol{\Psi}_j(t)$ in Example 3 results, e.g., in $k = 1$ and $\mathbf{Z}_n = 2\mathbf{I}$, $\mathbf{Z}_{n-1} = -\mathbf{I}$ since $\boldsymbol{\Psi}_j(T_{n+1}) = 2\boldsymbol{u}_j(T_n) - \boldsymbol{u}_j(T_{n-1})$. For constant extrapolation ($k = 0$) with $\boldsymbol{\Psi}_j(T_{n+1}) = \boldsymbol{u}_j(T_n)$, Jacobian $\mathbf{Z}_n$ is given by the identity matrix $\mathbf{I}$.

Estimate (11b) provides an error bound for $\|\boldsymbol{\epsilon}_{n+1}^{\boldsymbol{u}}\|$ in terms of $\mathcal{O}(1)\sum_\iota \|\boldsymbol{\epsilon}_{n-\iota}^{\boldsymbol{u}}\|$ that does in general not guarantee zero-stability and convergence of the modular time integration method [14]. There is a risk of exponential instability unless an additional contractivity condition is satisfied [4].

However, the $\sum_\iota \mathbf{J}_n\mathbf{Z}_{n-\iota}\boldsymbol{\epsilon}_{n-\iota}^{\boldsymbol{u}}$ term at the right hand side of (11b) may be eliminated by repeated application of this error estimate if condition (7) is satisfied for some finite $M > 0$. Substituting index n in (11b) by $n-\iota-1$, we get

$$\begin{aligned}\boldsymbol{\epsilon}_{n-\iota}^{\boldsymbol{u}} = \sum_{\kappa=0}^{k}\mathbf{J}_{n-\iota-1}\mathbf{Z}_{n-\iota-1-\kappa}\boldsymbol{\epsilon}_{n-\iota-1-\kappa}^{\boldsymbol{u}} + &\\ + \mathcal{O}(1)\|\boldsymbol{\epsilon}_{n-\iota-1}^{\boldsymbol{x}}\| + \mathcal{O}(H)\sum_{\kappa=0}^{k}\|\boldsymbol{\epsilon}_{n-\iota-1-\kappa}^{\boldsymbol{u}}\| + \mathcal{O}(H^{k+1})&\end{aligned} \tag{11c}$$

for all $\iota = 0, 1, \ldots, k$. This error recursion is inserted in the right hand side of (11b) resulting in

$$\begin{aligned}\boldsymbol{\epsilon}_{n+1}^{\boldsymbol{u}} = \sum_{\iota=0}^{k}\sum_{\kappa=0}^{k}\mathbf{J}_n\mathbf{Z}_{n-\iota}\mathbf{J}_{n-\iota-1}\mathbf{Z}_{n-\iota-1-\kappa}\boldsymbol{\epsilon}_{n-\iota-1-\kappa}^{\boldsymbol{u}} + &\\ + \mathcal{O}(1)\sum_{l=0}^{2k+1}\big(\|\boldsymbol{\epsilon}_{n-l}^{\boldsymbol{x}}\| + H\|\boldsymbol{\epsilon}_{n-l}^{\boldsymbol{u}}\|\big) + \mathcal{O}(H^{k+1})\,.&\end{aligned} \tag{11d}$$

For the coupled 1-DOF oscillators in Example 1, we have $\mathbf{J}_n = \mathbf{0}_{4\times4}$ in the case of displacement-displacement coupling and $\mathbf{J}_n \neq \mathbf{0}_{3\times3}$ but $\mathbf{J}_n\mathbf{Z}_{n-\iota}\mathbf{J}_{n-\iota-1}\mathbf{Z}_{n-\iota-1-\kappa} =$

$\mathbf{0}_{3\times 3}$ in the case of force-displacement coupling, see (8). For both settings, estimate (11d) simplifies to

$$\|\boldsymbol{\epsilon}^{\boldsymbol{u}}_{n+1}\| \le \overline{C}_u \sum_{l=0}^{M(k+1)-1} \left(\|\boldsymbol{\epsilon}^{\boldsymbol{x}}_{n-l}\| + H\,\|\boldsymbol{\epsilon}^{\boldsymbol{u}}_{n-l}\|\right) + \mathcal{O}(H^{k+1}) \tag{11e}$$

with $M = 2$. In the general case, estimate (11e) is obtained repeating the transformation steps from (11b) to (11d) another $M - 2$ times and exploiting the identity (7) that follows from the basic assumption that the coupled system (2) is represented by an acyclic directed graph G, see Sect. 2.

From the coupled error recursion (11a,e), we get global errors

$$\|\boldsymbol{\epsilon}^{\boldsymbol{x}}_{n}\| = \mathcal{O}(H^{k+1}), \quad \|\boldsymbol{\epsilon}^{\boldsymbol{u}}_{n}\| = \mathcal{O}(H^{k+1}),$$

i.e., convergence with order $k + 1$ on finite time intervals, see the classical convergence analysis for DAE time integration methods in [11], its application to modular time integration in [4] and the recent extension of these results to multistep signal extrapolation in [3].

This convergence result gives a theoretical justification for communication step size control algorithms being based on (estimates for) the local error since the global error on finite time intervals may be bounded in terms of the local error if the coupled system (2) is free of algebraic loops and condition (7) is satisfied for some finite $M > 0$. It shows furthermore that order reduction in the local error will not affect the order of the global error if there are no closed structural feed-through paths in the coupled system since by definition the length of all open structural feed-through paths is bounded by the dimension n_u of (5) such that $M \le n_u$ in (7).

4 Local Error Estimates

For communication step size control, an appropriate estimate of the local error is compared to user defined error bounds (tolerances), see [11]. Richardson extrapolation is a time-consuming but reliable way to estimate local errors in ODE and DAE time integration. Substantial savings of computing time result from an algorithmic modification that is tailored to the FMI co-simulation framework. In the present section both approaches are studied by an asymptotic error analysis. Numerical test results for a quarter car benchmark problem are presented in Sect. 5 below.

To keep the notation compact, we restrict the theoretical analysis of the error estimates to pure polynomial signal extrapolation in all subsystems that allows to perform and to analyse a communication step $T_n \to T_{n+1} = T_n + H$ for all subsystems in parallel. As in (9), we consider a polynomial $\overline{\boldsymbol{u}}(t)$ that interpolates the analytical solution $\boldsymbol{u}(t)$ of (2) at the $k + 1$ equidistant previous communication points $T_{n-\iota}$, $(\iota = 0, 1, \ldots, k)$. Classical estimates for the error of polynomial interpolation [9] prove

$$\boldsymbol{\Delta}_{\overline{u}}(t) := \overline{\boldsymbol{u}}(t) - \boldsymbol{u}(t) = -\frac{\boldsymbol{u}^{(k+1)}(T_n)}{(k+1)!} \prod_{\iota=0}^{k} (t - T_{n-\iota}) + \mathcal{O}(H^{k+2}) \tag{12}$$

for all $t \in [T_n, T_{n+2}]$, i.e., for all t in two subsequent communication steps of step size H. Replacing input function $\boldsymbol{u}(t)$ in (2) by $\overline{\boldsymbol{u}}(t)$, we get approximate solutions $\overline{\boldsymbol{x}}(t)$, $\overline{\boldsymbol{y}}(t)$ with $\boldsymbol{\Delta}_{\overline{x}}(T_n) = \mathbf{0}$ and

$$\left.\begin{aligned} \dot{\boldsymbol{\Delta}}_{\overline{x}}(t) &= \mathbf{A}_n\, \boldsymbol{\Delta}_{\overline{x}}(t) + \mathbf{B}_n\, \boldsymbol{\Delta}_{\overline{u}}(t) + \mathcal{O}(H^{k+2}) \\ \boldsymbol{\Delta}_{\overline{y}}(t) &= \mathbf{C}_n\, \boldsymbol{\Delta}_{\overline{x}}(t) + \mathbf{D}_n\, \boldsymbol{\Delta}_{\overline{u}}(t) + \mathcal{O}(H^{k+2}) \end{aligned}\right\} \tag{13}$$

for

$$\boldsymbol{\Delta}_{\overline{x}}(t) := \overline{\boldsymbol{x}}(t) - \boldsymbol{x}(t)\,, \quad \boldsymbol{\Delta}_{\overline{y}}(t) := \overline{\boldsymbol{y}}(t) - \boldsymbol{y}(t)\,, \quad (t \in [T_n, T_{n+2}])\,.$$

In (13), system matrices $\mathbf{A}_n$, $\mathbf{B}_n$, $\mathbf{C}_n$, $\mathbf{D}_n$ denote the Jacobians $\boldsymbol{f}_{\boldsymbol{x}}$, $\boldsymbol{f}_{\boldsymbol{u}}$, $\boldsymbol{g}_{\boldsymbol{x}}$, $\boldsymbol{g}_{\boldsymbol{u}}$ evaluated at $\boldsymbol{x} = \boldsymbol{x}(T_n)$, $\boldsymbol{u} = \boldsymbol{u}(T_n)$. The leading error term in $\boldsymbol{\Delta}_{\overline{x}}$ is obtained as solution of a linear time invariant system resulting in

$$\boldsymbol{\Delta}_{\overline{x}}(T_n + aH) = -\mathbf{B}_n \int_0^a \prod_{\iota=0}^{k} (\iota + s)\, \mathrm{d}s \cdot \frac{\boldsymbol{u}^{(k+1)}(T_n)}{(k+1)!} H^{k+2} + \mathcal{O}(H^{k+3})\,, \tag{14}$$

see also (12). In the next communication step $T_{n+1} \to T_{n+2} = T_n + 2H$, input function $\boldsymbol{u}(t)$ is substituted by a polynomial $\overline{\overline{\boldsymbol{u}}}(t)$ that interpolates

$$\boldsymbol{c}(\overline{\boldsymbol{y}}(t)) = \boldsymbol{c}(\boldsymbol{g}(\overline{\boldsymbol{x}}(t), \overline{\boldsymbol{u}}(t)))$$

at $t = T_{n+1}$ and $\overline{\boldsymbol{u}}(t) = \boldsymbol{u}(t) = \boldsymbol{c}(\boldsymbol{y}(t))$ at $t = T_{n+1-\iota}$, $(\iota = 1, \ldots, k)$. Equation (5) implies

$$\boldsymbol{u}(T_{n+1}) = \boldsymbol{c}\big(\boldsymbol{g}(\boldsymbol{x}(T_{n+1}), \boldsymbol{u}(T_{n+1}))\big)$$

and

$$\begin{aligned} \overline{\overline{\boldsymbol{u}}}(T_{n+1}) &= \overline{\boldsymbol{u}}(T_{n+1}) - \boldsymbol{\Delta}_{\overline{u}}(T_{n+1}) + \boldsymbol{c}\big(\boldsymbol{g}(\overline{\boldsymbol{x}}(T_{n+1}), \overline{\boldsymbol{u}}(T_{n+1}))\big) - \boldsymbol{c}\big(\boldsymbol{g}(\boldsymbol{x}(T_{n+1}), \boldsymbol{u}(T_{n+1}))\big) \\ &= \overline{\boldsymbol{u}}(T_{n+1}) - (\mathbf{I} - \mathbf{L}_n \mathbf{D}_n) \boldsymbol{\Delta}_{\overline{u}}(T_{n+1}) + \mathcal{O}(H^{k+2}) \end{aligned}$$

with $\mathbf{L}_n := (\partial \boldsymbol{c}/\partial \boldsymbol{y})(\boldsymbol{y}(T_n))$ since $\boldsymbol{\Delta}_{\overline{x}}(T_{n+1}) = \mathcal{O}(H^{k+2})$, see (14). According to (12), polynomial $\overline{\boldsymbol{u}}(t)$ approximates $\boldsymbol{u}(t)$ up to $\mathcal{O}(H^{k+1})$ and we get

$$\boldsymbol{\Delta}_{\overline{u}}(T_{n+1}) = -\boldsymbol{u}^{(k+1)}(T_n) \cdot H^{k+1} + \mathcal{O}(H^{k+2})\,.$$

The Lagrange form (9) of the interpolation polynomial shows

$$\bar{\bar{\boldsymbol{u}}}(t) - \bar{\boldsymbol{u}}(t) = \big(\bar{\bar{\boldsymbol{u}}}(T_{n+1}) - \bar{\boldsymbol{u}}(T_{n+1})\big) \prod_{l=1}^{k} \frac{t - T_{n+1-l}}{T_{n+1} - T_{n+1-l}}$$

$$= H(\mathbf{I} - \mathbf{L}_n\mathbf{D}_n)\frac{\boldsymbol{u}^{(k+1)}(T_n)}{(k+1)!}(k+1)\prod_{l=1}^{k}(t - T_{n+1-l}) + \mathcal{O}(H^{k+2})\,. \tag{15}$$

Since estimate (15) has the same basic structure as estimate (12), the *local error* at $t = T_{n+2}$

$$\boldsymbol{l}^{\boldsymbol{x}}_{n+2} := \boldsymbol{\Delta}_{\bar{\bar{\boldsymbol{x}}}}(T_{n+2}) = \boldsymbol{\Delta}_{\bar{\boldsymbol{x}}}(T_{n+2}) + \int_{T_{n+1}}^{T_{n+2}} \big(\boldsymbol{f}(\bar{\bar{\boldsymbol{x}}}(t), \bar{\bar{\boldsymbol{u}}}(t)) - \boldsymbol{f}(\bar{\boldsymbol{x}}(t), \bar{\boldsymbol{u}}(t))\big)\,\mathrm{d}t$$

may be expressed as

$$\boldsymbol{l}^{\boldsymbol{x}}_{n+2} = -\mathbf{B}_n(c_k\mathbf{I} + d_k\mathbf{L}_n\mathbf{D}_n)\frac{\boldsymbol{u}^{(k+1)}(T_n)}{(k+1)!}H^{k+2} + \mathcal{O}(H^{k+3}) \tag{16}$$

with constants

$$d_k := (k+1)\int_1^2 \prod_{\iota=1}^{k}(\iota + s - 1)\,\mathrm{d}s\,, \tag{17a}$$

$$c_k := \int_0^2 \prod_{\iota=0}^{k}(\iota + s)\,\mathrm{d}s - d_k = 2\int_0^1 \prod_{\iota=0}^{k}(\iota + s)\,\mathrm{d}s\,. \tag{17b}$$

For coupled systems with direct feed-through in some of the subsystems, we have $\mathbf{D}_n \neq \mathbf{0}$ and the local error $\boldsymbol{l}^{\boldsymbol{y}}_{n+2}$ in the output variables $\boldsymbol{y}$ is dominated by the error in components $\boldsymbol{u}$ at $t = T_{n+2}$:

$$\boldsymbol{l}^{\boldsymbol{u}}_{n+2} := \boldsymbol{\Delta}_{\bar{\bar{\boldsymbol{u}}}}(T_{n+2}) = \bar{\bar{\boldsymbol{u}}}(T_{n+2}) - \boldsymbol{u}(T_{n+2}) = \bar{\bar{\boldsymbol{u}}}(T_{n+2}) - \bar{\boldsymbol{u}}(T_{n+2}) + \bar{\boldsymbol{u}}(T_{n+2}) - \boldsymbol{u}(T_{n+2})$$

$$= -(\mathbf{I} + (k+1)\mathbf{L}_n\mathbf{D}_n)\boldsymbol{u}^{(k+1)}(T_n)\cdot H^{k+1} + \mathcal{O}(H^{k+2})\,, \tag{18}$$

see (12) and (15). Summarizing (16) and (18), we end up with

$$\boldsymbol{l}^{\boldsymbol{y}}_{n+2} = \boldsymbol{\Delta}_{\bar{\bar{\boldsymbol{y}}}}(T_{n+2}) = \mathbf{D}_n\boldsymbol{\Delta}_{\bar{\bar{\boldsymbol{u}}}}(T_{n+2}) + \mathbf{C}_n\boldsymbol{\Delta}_{\bar{\bar{\boldsymbol{x}}}}(T_{n+2}) + \mathcal{O}(c_{\mathbf{D}}H^{k+2} + H^{k+3})$$

$$= -\big((k+1)!\,\mathbf{D}_n(\mathbf{I} + (k+1)\mathbf{L}_n\mathbf{D}_n) + \tag{19}$$

$$+ H\mathbf{C}_n\mathbf{B}_n(c_k\mathbf{I} + d_k\mathbf{L}_n\mathbf{D}_n)\big)\frac{\boldsymbol{u}^{(k+1)}(T_n)}{(k+1)!}H^{k+1} + \mathcal{O}(c_{\mathbf{D}}H^{k+2} + H^{k+3})\,.$$

Constant $c_{\mathbf{D}}$ is set to $c_{\mathbf{D}} = 0$ if $\partial \boldsymbol{g}/\partial \boldsymbol{u} \equiv \mathbf{0}$ and to $c_{\mathbf{D}} = 1$ otherwise. For systems without direct feed-through, see (4), we have $\mathbf{D}_n = \mathbf{0}$ and $c_{\mathbf{D}} = 0$ and the right hand side in (19) simplifies substantially:

$$\boldsymbol{l}^{y}_{n+2} = -c_k \mathbf{C}_n \mathbf{B}_n \frac{\boldsymbol{u}^{(k+1)}(T_n)}{(k+1)!} H^{k+2} + \mathcal{O}(H^{k+3})\,. \tag{20}$$

For error estimation by Richardson extrapolation, we consider in each time interval $[T_n, T_{n+2}]$ a second numerical solution $\tilde{\boldsymbol{x}}(t)$ with $\tilde{\boldsymbol{x}}(T_n) = \boldsymbol{x}(T_n)$ and an input function $\tilde{\boldsymbol{u}}(t)$ that is defined by the interpolation polynomial for data points $\big(T_{n-2\iota}, \boldsymbol{u}(T_{n-2\iota})\big)$ with $T_{n-2\iota} = T_n - 2\iota H$, $(\iota = 0, 1, \ldots, k)$. As before, the interpolation error $\boldsymbol{\Delta}_{\tilde{\boldsymbol{u}}}(t)$ introduces errors $\boldsymbol{\Delta}_{\tilde{\boldsymbol{x}}}(t)$ and $\boldsymbol{\Delta}_{\tilde{\boldsymbol{y}}}(t)$ in state and output variables that may be studied by a perturbation analysis. We get

$$\boldsymbol{\Delta}_{\tilde{\boldsymbol{x}}}(T_{n+2}) = -\tilde{c}_k \mathbf{B}_n \frac{\boldsymbol{u}^{(k+1)}(T_n)}{(k+1)!} H^{k+2} + \mathcal{O}(H^{k+3}) \tag{21}$$

with

$$\tilde{c}_k := \int_0^2 \prod_{\iota=0}^{k} (2\iota + s)\, \mathrm{d}s = 2 \int_0^1 \prod_{\iota=0}^{k} (2\iota + 2\tilde{s})\, \mathrm{d}\tilde{s} = 2^{k+1} c_k\,,$$

see (17). Similar to estimate (12), the interpolation error in components $\boldsymbol{u}$ at $t = T_{n+2}$ is given by

$$\begin{aligned} \boldsymbol{\Delta}_{\tilde{\boldsymbol{u}}}(T_{n+2}) &= -\frac{\boldsymbol{u}^{(k+1)}(T_n)}{(k+1)!} \prod_{\iota=0}^{k} (T_{n+2} - T_{n-2\iota}) + \mathcal{O}(H^{k+2}) \\ &= -2^{k+1} \boldsymbol{u}^{(k+1)}(T_n) H^{k+1} + \mathcal{O}(H^{k+2})\,. \end{aligned} \tag{22}$$

Estimates (21) and (22) yield an error bound for the output variables:

$$\begin{aligned} \boldsymbol{\Delta}_{\tilde{\boldsymbol{y}}}(T_{n+2}) &= \mathbf{D}_n \boldsymbol{\Delta}_{\tilde{\boldsymbol{u}}}(T_{n+2}) + \mathbf{C}_n \boldsymbol{\Delta}_{\tilde{\boldsymbol{x}}}(T_{n+2}) + \mathcal{O}(c_{\mathbf{D}} H^{k+2} + H^{k+3}) \\ &= -2^{k+1} \big((k+1)!\, \mathbf{D}_n + H c_k \mathbf{C}_n \mathbf{B}_n \big) \frac{\boldsymbol{u}^{(k+1)}(T_n)}{(k+1)!} H^{k+1} + \\ &\qquad + \mathcal{O}(c_{\mathbf{D}} H^{k+2} + H^{k+3})\,. \end{aligned} \tag{23}$$

In ODE and DAE time integration, the comparison of the numerical results for a double-step with (small) step size H and a single (large) step with step size $2H$ allows to estimate precisely the leading error term of the local error [11]. For modular time integration, this error estimate is given by [13]

$$\mathbf{EST}_{\text{Rich}} := \frac{\tilde{\boldsymbol{y}}(T_{n+2}) - \overline{\overline{\boldsymbol{y}}}(T_{n+2})}{2^{k+1} - 1} = \frac{\boldsymbol{\Delta}_{\tilde{\boldsymbol{y}}}(T_{n+2}) - \boldsymbol{\Delta}_{\overline{\overline{\boldsymbol{y}}}}(T_{n+2})}{2^{k+1} - 1}\,. \tag{24}$$

The comparison of (19), (23) and (24) shows

$$\mathbf{EST}_{\mathrm{Rich}} = \boldsymbol{l}^{y}_{n+2} + \frac{2^{k+1}(k+1)}{2^{k+1}-1}\mathbf{D}_n\mathbf{L}_n\mathbf{D}_n\boldsymbol{u}^{(k+1)}(T_n)H^{k+1} + \\ + \mathscr{O}(c_{\mathbf{D}}H^{k+2} + H^{k+3})\,. \tag{25}$$

There is a qualitative difference to classical ODE and DAE theory: In the context of co-simulation, Richardson extrapolation may give asymptotically wrong results if $\mathbf{L}_n\mathbf{D}_n \neq \mathbf{0}$, i.e., for coupled systems with direct feed-through in at least one subsystem. If there are no subsystems with direct feed-through and condition (4) is satisfied, $\mathbf{EST}_{\mathrm{Rich}}$ reproduces, however, all components of the local error in the output variables correctly up to higher order terms. Therefore, this error estimate is considered to be one of the most reliable ways to approximate local errors in modular time integration.

In the ITEA2 project MODELISAR, several modifications of error estimate (24) were tested (Olsson, Private communication, June 2011) to reduce the large extra effort for computing $\tilde{\boldsymbol{y}}(t)$. Neglecting all details of practical implementation, the intermediate results $\overline{\boldsymbol{x}}(T_{n+1})$ and $\tilde{\boldsymbol{x}}(T_{n+1})$ coincide for constant extrapolation ($k = 0$) of input function $\boldsymbol{u}(t)$ since $\overline{\boldsymbol{u}}(t) = \tilde{\boldsymbol{u}}(t) = \boldsymbol{u}(T_n)$, ($t \in [T_n, T_{n+1}]$), in that case. From the view point of numerical efficiency, it would be favourable to use also for higher order extrapolation ($k \geq 1$) one and the same approximation $\overline{\boldsymbol{u}}(t)$ of the input function $\boldsymbol{u}(t)$ for both numerical solutions in the first communication step $T_n \to T_{n+1} = T_n + H$ and to restrict the use of different input functions to the second communication step, i.e., to $t \in [T_{n+1}, T_{n+2}]$.

In that way, co-simulation may proceed with a large communication step $T_n \to T_{n+2} = T_n + 2H$ of size $2H$ that is temporarily interrupted at $t = T_{n+1}$ to provide input data $\overline{\boldsymbol{y}}(T_{n+1})$ and $\boldsymbol{c}(\overline{\boldsymbol{y}}(T_{n+1}))$ for the second numerical solution to be used for error estimation. Alternatively, a small communication step $T_n \to T_{n+1} = T_n + H$ may be completed in the classical way and the two different numerical solutions on time interval $[T_{n+1}, T_{n+2}]$ are evaluated in parallel. With this second strategy, no subsystem solver has to go backward in time and the practical implementation might be simplified. Note, that both strategies are analytically equivalent but give slightly different results in a practical implementation since the internal (micro) step size sequences in the subsystems depend on the length of the communication steps.

With the notations of this section, the modified error estimate is given by

$$\mathbf{EST}_{\mathrm{mod}} := \frac{\overline{\boldsymbol{y}}(T_{n+2}) - \overline{\overline{\boldsymbol{y}}}(T_{n+2})}{c_{k,\mathrm{mod}} - 1} = \frac{\boldsymbol{\Delta}_{\overline{y}}(T_{n+2}) - \boldsymbol{\Delta}_{\overline{\overline{y}}}(T_{n+2})}{c_{k,\mathrm{mod}} - 1} \tag{26}$$

with some suitable constant $c_{k,\mathrm{mod}}$. To adjust this constant, we observe

$$\boldsymbol{\Delta}_{\overline{x}}(T_{n+2}) = -\overline{c}_k\mathbf{B}_n\frac{\boldsymbol{u}^{(k+1)}(T_n)}{(k+1)!}H^{k+2} + \mathscr{O}(H^{k+3}) \tag{27}$$

with

$$\overline{c}_k := \int_0^2 \prod_{\iota=0}^{k} (\iota + s)\, \mathrm{d}s$$

and set $c_{k,\text{mod}} := \overline{c}_k / c_k$ to get an asymptotically correct error estimate for systems without direct feed-through, see (20). For constant extrapolation, we have $c_{0,\text{mod}} = 2 = 2^{k+1}$ since $\overline{c}_0 = \tilde{c}_0$. For linear and quadratic extrapolation, the constants $c_{1,\text{mod}} = 14/5$ and $c_{2,\text{mod}} = 32/9$ are used.

For coupled systems with direct feed-through in one of the subsystems, we consider the error in input function $\overline{\boldsymbol{u}}(t)$ at $t = T_{n+2}$:

$$\begin{aligned}
\boldsymbol{\Delta}_{\overline{u}}(T_{n+2}) &= -\frac{\boldsymbol{u}^{(k+1)}(T_n)}{(k+1)!} \prod_{\iota=0}^{k} (T_{n+2} - T_{n-\iota}) + \mathcal{O}(H^{k+2}) \\
&= -(k+2)\boldsymbol{u}^{(k+1)}(T_n) H^{k+1} + \mathcal{O}(H^{k+2}) \, .
\end{aligned}$$

This error is $k+2$ times larger than the first error term in $\boldsymbol{l}^{\boldsymbol{u}}_{n+2}$, see (18), implying the necessary condition $c_{k,\text{mod}} = k+2$ for an asymptotically correct modified error estimate (26) in the case of direct feed-through in one of the subsystems ($\mathbf{D}_n \neq \mathbf{0}$). This condition is trivially satisfied for constant extrapolation but slightly violated in the linear and quadratic case since $(1+2) - c_{1,\text{mod}} = 3 - 14/5 = 1/5$ and $(2+2) - c_{2,\text{mod}} = 4 - 32/9 = 4/9$.

5 Numerical Test: Benchmark Quarter Car

Figure 3 shows the strongly simplified model of a quarter car with two point masses m_w and m_c representing the primary mass (unsprung mass) and the secondary mass (spring mounted mass), respectively, see [16]. The point masses have a vertical

$$\begin{aligned}
m_\text{c}\ddot{x}_\text{c} &= F_\text{susp}(x_\text{c}, \dot{x}_\text{c}, x_\text{w}, \dot{x}_\text{w}) \\
m_\text{w}\ddot{x}_\text{w} &= F_\text{tire}(x_\text{w}, \dot{x}_\text{w}, u_\text{ex}(t)) - F_\text{susp}(x_\text{c}, \dot{x}_\text{c}, x_\text{w}, \dot{x}_\text{w})
\end{aligned}$$

$$u_\text{ex}(t) = (z(t), \dot{z}(t))^\top, \; z(t) = \begin{cases} 0 & \text{if } t \le 0 \\ 0.1 & \text{if } t > 0 \end{cases}$$

$$\begin{aligned}
F_\text{susp} &= k_\text{c}(x_\text{w} - x_\text{c}) + d_\text{c}(\dot{x}_\text{w} - \dot{x}_\text{c}) \\
F_\text{tire} &= k_\text{w}(z - x_\text{w}) + d_\text{w}(\dot{z} - \dot{x}_\text{w})
\end{aligned}$$

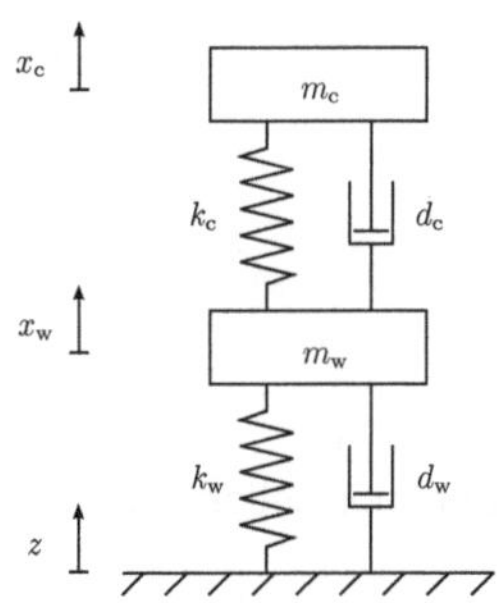

Fig. 3 Benchmark Quarter car: equations of motion

degree of freedom. They are coupled by a linear spring-damper element F_{susp} representing the secondary suspension. Tire forces are considered by an additional spring-damper element F_{tire} between unsprung mass and ground.

The road profile is considered as external system input $z(t)$. At $t = 0$ the system starts in its equilibrium state $x_{\mathrm{c}} = x_{\mathrm{w}} = 0$ and is immediately excited by the jump discontinuity of z. For model parameters according to Knorr [12], Fig. 4 shows the vertical displacements illustrating the different time scales in unsprung and spring mounted mass, respectively.

In co-simulation, both subsystems are integrated separately by standard ODE integrators with very fine error tolerances. The subsystems are either coupled by a displacement-displacement coupling or by a force-displacement coupling, see Example 1. For various values of the communication step size H, Figs. 5 and 6 compare the local error $\|\boldsymbol{l}^{y}_{n+2}\|_2$ of two consecutive communication steps $T_n \to T_n + H$, $T_n + H \to T_n + 2H$, see (19), with the error estimates $\mathbf{EST}_{\mathrm{mod}}$ and $\mathbf{EST}_{\mathrm{Rich}}$. Note, that the absolute values of $\boldsymbol{l}^{y}_{n+2}$ are substantially larger in the case of force-displacement coupling (Fig. 6) since $\boldsymbol{u}_1 = \boldsymbol{y}_2 = F_{\mathrm{susp}}$ in that case, see Example 1. For displacement-displacement coupling (Fig. 5), the components of $\boldsymbol{y}$ are given by position and velocity coordinates x_{c}, $\dot{x}_{\mathrm{c}}$, x_{w}, $\dot{x}_{\mathrm{w}}$ that are of much smaller magnitude than $|F_{\mathrm{susp}}|$. Furthermore, the numerical test results in

m_{c}	400.0 kg
m_{w}	40.0 kg
k_{c}	15000.0 N/m
k_{w}	150000.0 N/m
d_{c}	1000.0 Ns/m
d_{w}	0.0 Ns/m

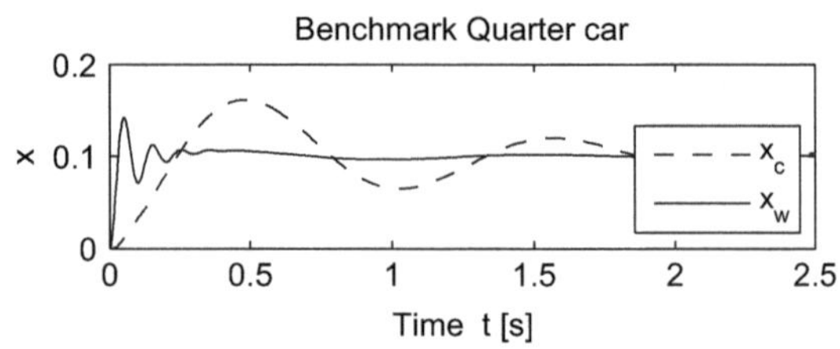

Fig. 4 Benchmark Quarter car: model parameters and solution [12]

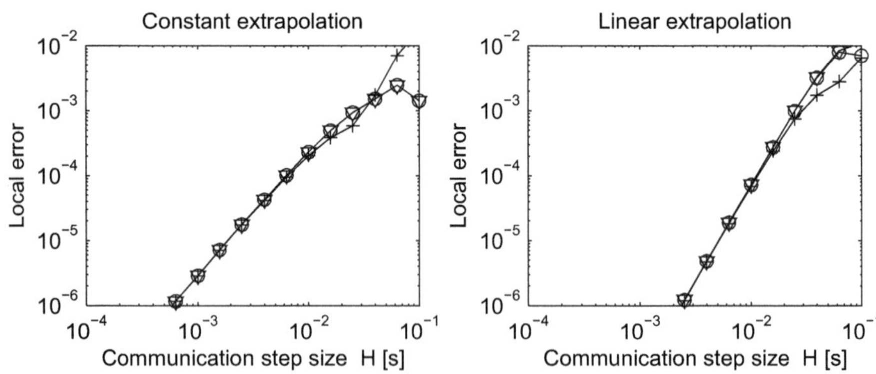

Fig. 5 Benchmark Quarter car, co-simulation with displacement-displacement coupling: local error ("+") and error estimates $\mathbf{EST}_{\mathrm{Rich}}$, $\mathbf{EST}_{\mathrm{mod}}$ ("∘", "∇")

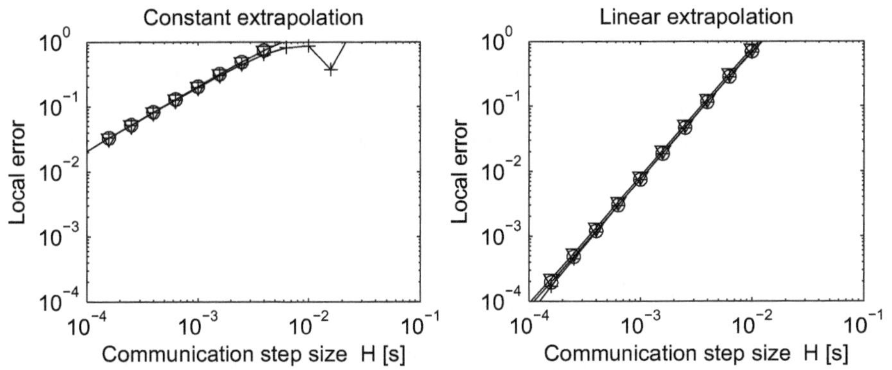

Fig. 6 Benchmark Quarter car, co-simulation with force-displacement coupling: local error ("+") and error estimates $\mathbf{EST}_{\text{Rich}}$, $\mathbf{EST}_{\text{mod}}$ ("∘", "∇")

Figs. 5 and 6 illustrate $\|\boldsymbol{l}^y_{n+2}\|_2 = \mathcal{O}(H^{k+2})$ for systems without direct feed-through (displacement-displacement coupling, $c_{\mathbf{D}} = 0$, Fig. 5) and the reduced order $\|\boldsymbol{l}^y_{n+2}\|_2 = \mathcal{O}(H^{k+1})$ for systems with direct feed-through in at least one subsystem (force-displacement coupling, $c_{\mathbf{D}} = 1$, Fig. 6), see (19) and (20), respectively. In both cases, the results are shown for $k = 0$ (constant interpolation) and for $k = 1$ (linear interpolation). The new, alternative error estimate $\mathbf{EST}_{\text{mod}}$ is as reliable as the classical estimate $\mathbf{EST}_{\text{Rich}}$.

The local error analysis in Sect. 4 and numerical test results like the ones in Figs. 5 and 6 are important prerequisites for a theoretically justified communication step size control. The end user, however, is not interested in *local* errors but in *global* ones. The convergence analysis in Sect. 3 shows that neither for displacement-displacement coupling nor for force-displacement coupling the global errors $\boldsymbol{\epsilon}^x_n$, $\boldsymbol{\epsilon}^y_n$, $\boldsymbol{\epsilon}^u_n$ suffer from order reduction. Moreover, the global errors $\boldsymbol{\epsilon}^x_n$ are expected to be in the same order of magnitude for both types of coupling since the state vector $\boldsymbol{x}$ has in both cases the same components x_{c}, $\dot{x}_{\text{c}}$, x_{w} and $\dot{x}_{\text{w}}$.

Note, that for a practical implementation of local error estimates like $\mathbf{EST}_{\text{Rich}}$ and $\mathbf{EST}_{\text{mod}}$ in communication step size control, the discretization errors of the (micro) time integration in the individual subsystems can no longer be neglected. This open problem is subject of future research, see also [18] for some preliminary results.

Conclusions

The numerical efficiency of co-simulation algorithms may be improved substantially by higher order approximations of subsystem inputs and by variable communication step sizes. A strict mathematical analysis shows that the global error is bounded in terms of local errors if there are no algebraic loops in the coupled system. Local error estimates based on Richardson

(continued)

extrapolation techniques and some modifications have been studied in detail proving their efficiency in systems without direct feed-through like mechanical systems with displacement-displacement coupling. A force-displacement coupling results in direct feed-through in one of the subsystems and may deteriorate the favourable asymptotic properties of classical error estimation strategies.

Acknowledgements The authors are grateful to the Polish Academy of Sciences for granting the permission to reproduce this copyrighted material from *Archive of Mechanical Engineering* **LX**(2013)75–94 in this proceedings volume. They furthermore gratefully acknowledge the fruitful cooperation in the European ITEA2 project "MODELISAR - From system modelling to S/W running on the vehicle" (2008–2011) and valuable discussions with J. Bastian, T. Blochwitz (Dresden), H. Elmqvist, H. Olsson (Lund), M. Otter (Oberpfaffenhofen) and other partners of the MODELISAR consortium. The work was supported by the German Minister of Education and Research, BMBF projects 01IS08002N and 03MS633A. The research on the convergence analysis for modular time integration methods applied to coupled systems in block representation was inspired by fruitful discussions with M. Busch (Kassel).

References

1. Arnold, M.: Multi-rate time integration for large scale multibody system models. In: Eberhard, P. (ed.) IUTAM Symposium on Multiscale Problems in Multibody System Contacts, Stuttgart, pp. 1–10. Springer (2007)
2. Arnold, M.: Numerical methods for simulation in applied dynamics. In: Arnold, M., Schiehlen, W. (eds.) Simulation Techniques for Applied Dynamics. Volume 507 of CISM Courses and Lectures, pp 191–246. Springer, Wien/New York (2009)
3. Arnold, M.: Stability of sequential modular time integration methods for coupled multibody system models. J. Comput. Nonlinear Dyn. **5**, 031003 (2010)
4. Arnold, M., Günther, M.: Preconditioned dynamic iteration for coupled differential-algebraic systems. BIT Numer. Math. **41**, 1–25 (2001)
5. Busch, M.: Zur effizienten Kopplung von Simulationsprogrammen. PhD thesis, Universität Kassel, Fachbereich Maschinenbau (2012)
6. Busch, M., Schweizer, B.: Numerical stability and accuracy of different co-simulation techniques: analytical investigations based on a 2-DOF test model. In: Proceedings of the 1st Joint International Conference on Multibody System Dynamics, Lappeenranta, 25–27 May 2010
7. Clauß, C., Arnold, M., Schierz, T., Bastian, J.: Master zur Simulatorkopplung via FMI. In: Liu-Henke, X. (ed.) Tagungsband der ASIM/GI-Fachgruppen STS und GMMS, Wolfenbüttel, 23–24 Feb 2012. Ostfalia Hochschule für Angewandte Wissenschaften, Wolfenbüttel (2012)
8. Cormen, T.H., Leiserson, C.E., Rivest, R.L., Stein, C.: Introduction to Algorithms, 2nd edn. MIT, Cambridge (2001)
9. Deuflhard, P., Hohmann, A.: Numerical Analysis in Modern Scientific Computing: An Introduction. Number 43 in Texts in Applied Mathematics, 2nd edn. Springer, New York (2003)
10. FMI: The Functional Mockup Interface. https://www.fmi-standard.org/
11. Hairer, E., Wanner, G.: Solving Ordinary Differential Equations. II. Stiff and Differential-Algebraic Problems, 2nd edn. Springer, Berlin/Heidelberg/New York (1996)

12. Knorr, S.: Multirate-Verfahren in der Co-Simulation gekoppelter dynamischer Systeme mit Anwendung in der Fahrzeugdynamik. Master thesis, University Ulm, Faculty of Mathematics and Economics (2002)
13. Kübler, R.: Modulare Modellierung und Simulation mechatronischer Systeme. Fortschritt-Berichte VDI Reihe 20, Nr. 327. VDI–Verlag GmbH, Düsseldorf (2000)
14. Kübler, R., Schiehlen, W.: Two methods of simulator coupling. Math. Comput. Model. Dyn. Syst. **6**, 93–113 (2000)
15. Modelica Association Project FMI: Functional Mockup Interface for Model Exchange and Co-Simulation v2.0 beta 4. https://www.fmi-standard.org/downloads (2012)
16. Popp, P., Schiehlen, W.O.: Ground Vehicle Dynamics. Springer, Berlin/Heidelberg (2010)
17. Schierz, T., Arnold, M.: Stabilized overlapping modular time integration of coupled differential-algebraic equations. Appl. Numer. Math. **62**, 1491–1502 (2012)
18. Schierz, T., Arnold, M., Clauß, C.: Co-simulation with communication step size control in an FMI compatible master algorithm. In: Otter, M., Zimmer, D. (eds.) Proceedings of the 9th International Modelica Conference, Munich, 3–5 Sept 2012
19. Walter, W.: Ordinary Differential Equations. Number 182 in Graduate Texts in Mathematics. Springer, New York (1998)

A Unified (P)DAE Modeling Approach for Flow Networks

Lennart Jansen and Caren Tischendorf

Abstract We present a unified modeling approach for different types of flow networks, for instance electric circuits, water and gas supplying networks. In all cases the flow network is described by the pressures at the nodes of the network and the flows through the branches of the network. It is shown that the mass balance equations at each node are independent of the type of flow medium and can be described by the use of incidence matrices reflecting the network topology. Additionally, various types of net element models are presented. Finally, all network describing equations are summarized for some prototype networks which differ by the various net element models. They yield in pure linear/nonlinear equation systems, differential-algebraic systems or partial differential equation systems. All of them may have serious rank changes in the model functions if switching elements belong to the network. The model descriptions presented here keep all the network structure information and can be exploited for the analysis, numerical simulation and optimization of such networks.

Keywords Modeling • Flow network • Partial differential algebraic equation • Circuit • Water network • Gas network

Mathematics Subject Classification (2010) 93A30 · 34A09 · 35M20 · 94C05 · 34B45 · 76N15

1 Modeling of Flow Networks

In flow networks, a certain medium is flowing through branches that are connected by nodes. The medium can be very different, for example, gas, water and electric currents. Pressure or potential differences between nodes cause a flow through the connecting branches of the network. In order to describe the flows through the

---

L. Jansen • C. Tischendorf (✉)
Humboldt-Univerität zu Berlin, 10099 Berlin, Germany,
e-mail: lejansen@math.hu-berlin.de; tischendorf@math.hu-berlin.de

S. Schöps et al. (eds.), *Progress in Differential-Algebraic Equations*,
Differential-Algebraic Equations Forum, DOI 10.1007/978-3-662-44926-4_7

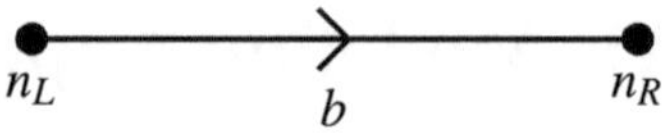

Fig. 1 Network branch b directing from left node n_L to right node n_R

branches and the corresponding pressures or potentials at each node, we consider the network as an oriented graph $G = (N, B)$ with $N = \{n_1, n_2, \ldots, n_N\}$ forming the set of all nodes and $B = \{b_1, b_2, \ldots, b_B\}$ being the set of all branches. We equip each branch with a certain direction. It allows us to distinguish between the two nodes of each branch: a left node n_l and a right node n_r. Each branch directs from the left to the right node (see Fig. 1).

Remark 1 The direction of a branch from the left to the right node does not mean that the medium is always flowing from the left to the right node. Later on, we see that the branch direction tells us how we have to interpret the sign of the flow values. If the flow has a positive sign then it flows from left to right. If the flow has a negative sign then it flows from right to left.

The assignment of the left and right nodes of each branch to the global node numbers may easily described by the incidence matrices $A_L, A_R \in \mathbb{R}^{n_N \times n_B}$ defined by

$$(A_L)_{ij} = \begin{cases} -1 & \text{if node } n_i \text{ is the left node of branch } b_j, \\ 0 & \text{else,} \end{cases}$$

$$(A_R)_{ij} = \begin{cases} +1 & \text{if node } n_i \text{ is the right node of branch } b_j, \\ 0 & \text{else.} \end{cases}$$

For later use, we additionally introduce the incidence matrix

$$A := A_L + A_R \in \mathbb{R}^{n_N \times n_B}.$$

Obviously, we have

$$(A)_{ij} = \begin{cases} -1 & \text{if node } n_i \text{ is the left node of branch } b_j, \\ +1 & \text{if node } n_i \text{ is the right node of branch } b_j, \\ 0 & \text{else.} \end{cases} \tag{1}$$

Remark 2 We exclude in our considerations networks with self-loops. With other words, the network is not allowed to have branches with left and right node being the same node. It ensures the correctness of Eq. (1).

Remark 3 The definition of the incidence matrices differs in the various communities analyzing flow networks. For water supplying networks, one usually uses

$$A_{\text{water}} = A^{\top} \in \mathbb{R}^{n_B \times n_N},$$

see e.g. [13, 31, 34] but in electric circuit analysis one commonly uses

$$A_{\text{circuit}} = -A \in \mathbb{R}^{n_N \times n_B},$$

see e.g. [9, 12, 16].

1.1 Network Topology Describing Equations

Next, we describe the mass flow balance of each flow network. Independently of the kind of the flow medium, the sum of all flows entering one node equals the sum of all flows leaving this node. It reflects the law of conservation of mass at each node. In circuit analysis, it is known as the Kirchhoff's current law.

Introducing the flow variables $q_L, q_R \in \mathbb{R}^{n_B}$ as vectors with the entries

$$(q_L)_i = \text{flow entering branch } b_i \text{ at the left node,}$$
$$(q_R)_i = \text{flow leaving branch } b_i \text{ at the right node,}$$

we may write the mass flow balance equations as

$$A_L q_L + A_R q_R = 0. \tag{2}$$

The i-the row of this equation system reflects the mass flow balance equation at node n_i. If the network contains nodes connected to flow sources (e.g. to reservoirs) or to flow sinks (e.g. to open hydrants) then we rewrite the mass flow balance equations (2) as

$$A_L q_L + A_R q_R = q_s \tag{3}$$

with

$$(q_s)_j = \text{sum of flows entering/leaving node } n_j \text{ from sources/sinks}$$

for all $j = 1, \ldots, n_N$.

Remark 4 If we may neglect the time delay the flow impulse needs to be transferred from the left to the right node of a branch (it depends on the flow medium and the flow distances) then we may identify q_L with q_R and operate only with

$$q := q_L = q_R.$$

The mass flow balance equations (3) then reduce to

$$Aq = q_s. \tag{4}$$

The time delay is, for example, neglected in standard circuit analysis. However, one can not neglect it, for instance, for the gas flow through longer pipe lines or for the current flow through transmission lines.

Remark 5 The mass flow q is denoted differently for the different flow media. We commonly find the following notations in the literature:

$$
\begin{aligned}
q_{\text{water}} &= m \text{ or } Q && \text{water flow,} \\
q_{\text{gas}} &= q && \text{gas flow,} \\
q_{\text{circuit}} &= i && \text{current flow,} \\
q_{\text{blood}} &= Q && \text{blood flow.}
\end{aligned}
$$

As already mentioned, pressure or potential differences between two nodes cause a flow through the branches connecting these nodes. Let $p \in \mathbb{R}^{n_N}$ denote the vector of pressure/potential at the nodes, that means, $p_i \in \mathbb{R}$ describes the pressure/potential at the node n_i for $i = 1, \ldots, n_N$.

Remark 6 The pressure/potential notation differs in the literature depending on the type of flow media:

$$
\begin{aligned}
p_{\text{water}} &= h \text{ or } H && \text{(scaled) water pressure,} \\
p_{\text{gas}} &= p && \text{gas pressure,} \\
p_{\text{circuit}} &= e && \text{electric potentials,} \\
p_{\text{blood}} &= p && \text{blood pressure.}
\end{aligned}
$$

The incidence matrix A allows us easily to describe the pressure/potential differences $(\Delta p)_i$ at each branch b_i of the network. We have

$$(\Delta p)_i = p_{R(i)} - p_{L(i)} \qquad \forall i = 1, \ldots, n_B \tag{5}$$

for $L(i)$ and $R(i)$ being the left and right node for the branch b_i. Regarding the entries of A, we see that (5) is equivalent to

$$\Delta p = A^\top p \tag{6}$$

where the i-th component of Δp represents the pressure/potential differences of the i-th branch in the network. Notice that we have

$$\Delta p = A_R^\top p + A_L^\top p = p_R - p_L$$

with

$$p_R := A_R^\top p, \qquad p_L := -A_L^\top p. \tag{7}$$

The i-th component of p_R and p_L represent the pressure $p_{R(i)}$ of the right node $n_{R(i)}$ of pipe i and the pressure $p_{L(i)}$ of the left node $n_{L(i)}$ of pipe i.

We summarize the mass flow balance equation (3) and the pressure difference equation (6).

$$A_R q_R + A_L q_L = q_s, \tag{8}$$

$$\Delta p = A^\top p = A_R^\top p + A_L^\top p = p_R - p_L. \tag{9}$$

They form the network modeling equations describing the network topology. Next, we need the network equations describing the relations between the flows q and the pressures p or pressure differences Δp. These relations are element dependent and, therefore, differ for the various type of flow networks.

1.2 Network Element Modeling

We distinguish between two types of elements: branch elements and node elements. Node elements describe the pressure $\tilde{p}$ at a node $\tilde{n}$. Branch elements describe the relations between the flow $\tilde{q}_L$ and the pressure $\tilde{p}_L$ at the left node of a branch and the flow $\tilde{q}_R$ and the pressure $\tilde{p}_R$ at the right node.

If the node of a node element has the global node number j then we have

$$\tilde{p} = p_j.$$

If the branch of a branch element has the global branch number i then we have

$$\begin{aligned}
\tilde{q}_L &= (q_L)_i, \\
\tilde{q}_R &= (q_R)_i, \\
\tilde{p}_L &= (p_L)_i = (-A_L^\top p)_i, \\
\tilde{p}_R &= (p_R)_i = (A_R^\top p)_i.
\end{aligned}$$

For convenience, we neglect the tilde notation in the following. The reader should have in mind that each scalar p, p_L, p_R, q_L and q_R (Fig. 2) in the following reflects one component of the vectors p, p_L, p_R, q_L and q_R introduced before.

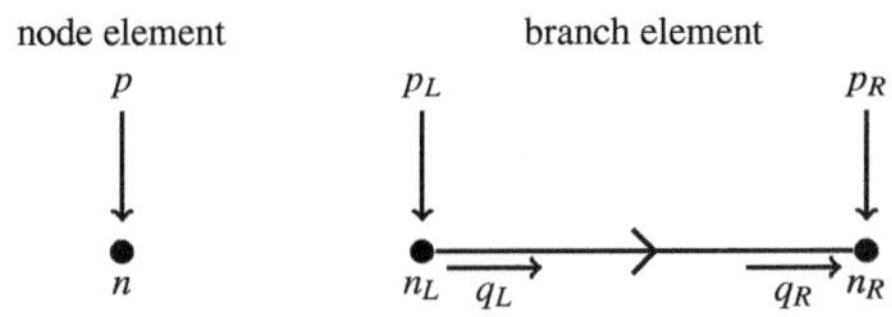

Fig. 2 Types of network elements: node elements and branch elements

Next, we collect some commonly used network element models for water networks, gas networks and electric circuits. The list of elements presented here is not complete but shall exemplify the variety of element models. Furthermore, we want to point out that these element models can also be used for the simulation of other flow networks. For instance, the blood flow can be modelled by circuit element models, see [11, 24].

Electric Circuits

For better understanding, we start with an example. Figure 3 represents a simple dynamic circuit for an induction machine. The rotor circuit part consists of a sinusoidal voltage source p_{s2} and a rotor resistance R_2. The stator circuit part has a constant voltage source p_{s1}, a stator resistance R_1, a leakage inductance L_1 and a magnetizing inductance L_2. For the flows of this lumped model we have

$$q_i = q_{Li} = q_{Ri} \qquad \forall i = 1, \ldots, 5.$$

The flow balance equations (8) are given by

$$\begin{aligned} q_1 - q_2 &= 0 \\ q_2 - q_3 &= 0 \\ q_3 - q_4 - q_5 &= 0 \\ q_5 - q_6 &= 0 \end{aligned}$$

The flow balance equation for the mass node can be neglected since it equals the negative sum of all other flow balance equations. The voltage difference equations (9) are represented by

$$\begin{aligned} (\Delta p)_1 &= p_1 - p_0 \\ (\Delta p)_2 &= p_2 - p_1 \\ (\Delta p)_3 &= p_3 - p_2 \end{aligned}$$

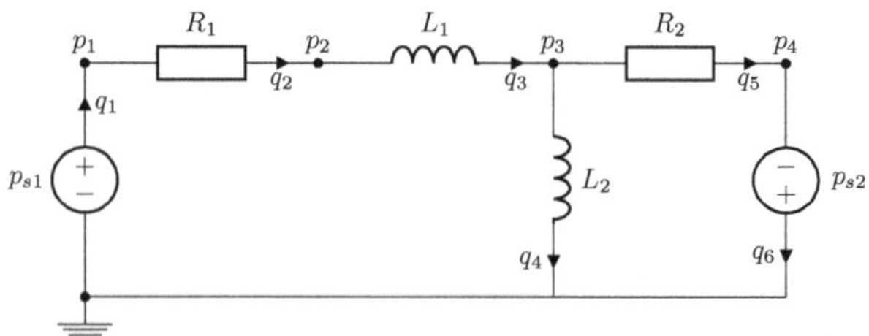

Fig. 3 Lumped inductor circuit with a static voltage source p_{s1} and a rotor voltage source p_{s2}

$$(\Delta p)_4 = p_0 - p_3$$
$$(\Delta p)_5 = p_4 - p_3$$
$$(\Delta p)_6 = p_0 - p_4$$

where the nodal voltage p_0 of the mass node is usually fixed by $p_0 = 0$. The net element models of the two voltage sources are given by

$$(\Delta p)_1 = p_{s1}$$
$$(\Delta p)_6 = p_{s2}.$$

The net element models of the two resistors are given by

$$(\Delta p)_2 = R_1 q_2$$
$$(\Delta p)_5 = R_2 q_5.$$

The net element models of the two inductors are given by

$$(\Delta p)_3 = L_1 q_3$$
$$(\Delta p)_4 = L_2 q_4.$$

In general, we have the following net element descriptions.

Resistor A resistor is a branch element. Ohmic resistors describe a linear relation between the voltage $p_L - p_R$ and the current $q = q_L = q_R$:

$$\Delta p = Rq$$

with the resistance $R > 0$. In general, there is a function f_{re} such that

$$\Delta p = f_{re}(q). \tag{10}$$

Conductor A conductor is a branch element. Ideal conductors describe a linear relation between the current $q = q_L = q_R$ and the voltage Δp:

$$q = G \cdot \Delta p$$

with the conductance $G > 0$. In general, there is a function f_{co} such that

$$q = f_{co}(\Delta p). \tag{11}$$

Capacitor A capacitor is a branch element. Its current $q = q_L = q_R$ is given by the time derivative of the charge. For ideal capacitors, the charge is given by $C \cdot \Delta p$ such that

$$q = C\frac{\mathrm{d}}{\mathrm{d}t}\Delta p$$

with the capacitance $C > 0$. In general, there is a function f_{ca} such that

$$q = \frac{\mathrm{d}}{\mathrm{d}t} f_{ca}(\Delta p). \tag{12}$$

Inductor An inductor is a branch element. Its voltage Δp is given by the time derivative of the magnetic flux. For ideal inductors, the flux can be described as $L \cdot q$ with the current $q = q_L = q_R$ such that

$$\Delta p = L\frac{\mathrm{d}}{\mathrm{d}t}q$$

with the inductance $L > 0$. In general, there is a function f_{in} such that

$$\Delta p = \frac{\mathrm{d}}{\mathrm{d}t} f_{in}(q). \tag{13}$$

Memristor A memristor is a branch element. Its current $q = q_L = q_R$ and its voltage Δp are related by the charge as follows, see [8, 25, 28]:

$$\Delta p = \frac{\mathrm{d}}{\mathrm{d}t} f_{mr}(u), \quad q = \frac{\mathrm{d}}{\mathrm{d}t}u$$

with a given function f_{mr} and an extra variable u reflecting the charge.

Memductor A memductor (also called flux-controlled memristor) is a branch element. Its current $q = q_L = q_R$ and its voltage Δp are related by the flux as follows, see [8, 27]:

$$q = \frac{\mathrm{d}}{\mathrm{d}t} f_{md}(u), \quad \Delta p = \frac{\mathrm{d}}{\mathrm{d}t}u \tag{14}$$

with a given function f_{md} and an extra variable u representing the flux. A modeling discussion of further memelements can be found, for example, in [1, 14, 26].

Voltage source A voltage source is a branch element. It prescribes the branch voltage Δp. An independent voltage source is represented by

$$\Delta p = f_{vs}(t) \tag{15}$$

with a given function f_{vs}. Typically, f_{vs} is either constant or a periodic function of time. Voltage sources can also be controlled ones. If a voltage source is current controlled then we have

$$\Delta p = f_{vs}(q_c) \tag{16}$$

with a given function f_{vs} and the controlling current q_c. If a voltage source is voltage controlled then we have

$$\Delta p = f_{vs}(\Delta p_c) \tag{17}$$

with a given function f_{vs} and the controlling voltage Δp_c.

Current source A current source is a branch element. It prescribes the current $q = q_L = q_R$. An independent current source is represented by

$$q = f_{cs}(t) \tag{18}$$

with a given function f_{cs}. Typically, f_{cs} is either constant or a periodic function of time. Current sources can be also controlled ones. If a current source is current controlled then we have

$$q = f_{cs}(q_c) \tag{19}$$

with a given function f_{cs} and the controlling current q_c. If a current source is voltage controlled then we have

$$q = f_{cs}(\Delta p_c) \tag{20}$$

with a given function f_{cs} and the controlling voltage Δp_c.

Diode A diode is a branch element. The Shockley ideal model description for a diode is given by

$$q = q_S(\exp^{\frac{\Delta p}{p_T}} - 1) =: f_{di}(\Delta p) \tag{21}$$

with q_S being the saturation current and

$$p_T = \frac{kT}{q_e}$$

denoting the thermal voltage, k the Boltzmann constant, T the temperature and q_e the elementary charge. The function f_{di} is monotone but not strongly monotone, see Fig. 4a. Considering real diodes, we have a characteristics as presented in Fig. 4b which can be described by a strongly monotone function f_{di} satisfying

$$q = f_{di}(\Delta p). \tag{22}$$

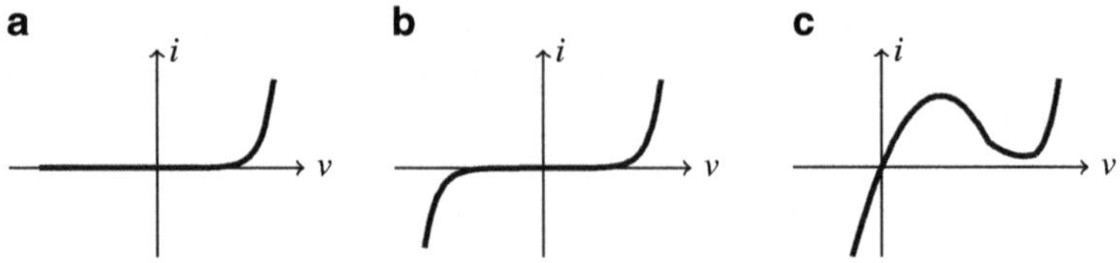

Fig. 4 Diode model functions. **(a)** Ideal Shockley diode. **(b)** Z-diode with a strongly monotone characteristics. **(c)** Tunnel diode

Strong monotonicity properties are crucial for the existence of unique global solutions [20]. However, there are also diodes that have a non-monotone characteristics, for instance tunnel diodes, see Fig. 4c.

Transistor A transistor is a multi-branch element. Typically, equivalent circuits with the basic circuits elements presented before are used to describe their behavior, see for instance [22]. These equivalent circuits become more and more complex. For instance, the BSIM4 model (see [6]) includes more than 800 parameter values. Not all of them have a direct physical interpretation. Therefore, it becomes more and more difficult to tune them for the various frequency regions in which the transistor is operating.

An alternative of such complex equivalent circuits is the use of discretized PDE models that describe the charge carrier movement and the electromagnetic field propagation. A commonly used model are the drift diffusion equations, see for instance [17,23]. For its embedding into network models see [4,29,32,33].

$$\frac{\mathrm{d}}{\mathrm{d}t}(Wp_c) - \int_{\Omega_S} g(x)(f_{Jn}(u_n, u_{\varphi S}, x) + f_{Jp}(u_p, u_{\varphi S}, x))\,\mathrm{d}x = q_c \tag{23a}$$

$$-\frac{\partial}{\partial t}u_n + \frac{1}{c_q}\nabla \cdot f_{Jn}(u_n, u_{\varphi S}, x) = f_R(u_n, u_p) \tag{23b}$$

$$\frac{\partial}{\partial t}u_p + \frac{1}{c_q}\nabla \cdot f_{Jp}(u_p, u_{\varphi S}, x) = -f_R(u_n, u_p) \tag{23c}$$

$$\nabla \cdot (-c_{\epsilon S}\nabla u_{\varphi S}) = c_q(p - n + c_C) \tag{23d}$$

$$\nabla \cdot (-c_{\epsilon O}\nabla u_{\varphi O}) = 0 \tag{23e}$$

with q_c being the flow entering the node with pressure p_c connected to the contact area Γ_c of the transistor. The matrix W is symmetric and positive definite. The extra variables are the space and time dependent electrostatic potentials $u_{\varphi S}$ and $u_{\varphi O}$ for the substrate and oxide region and the electron and hole densities u_n and u_p in the substrate region. The constants c_q, c_C, $c_{\epsilon S}$ and $c_{\epsilon O}$ are the elementary charge, the doping profile and the dielectric constants of the substrate and the oxide. The functions f_{Jn} and f_{Jp} describe the electron and hole current densities as

$$f_{Jn}(u_n, u_{\varphi S}, x) := q\mu_n(x)(c_{UT}\nabla u_n - u_n\nabla u_{\varphi S}),$$
$$f_{Jp}(u_p, u_{\varphi S}, x) := -q\mu_p(x)(c_{UT}\nabla u_p + u_p\nabla u_{\varphi S})$$

The functions μ_n and μ_p are given functions of x describing the electron and hole mobilities. The function f_R is a given function describing the balance of generation and recombination of electrons and holes. The drift diffusion equations are complete with the Dirichlet boundary conditions

$$u_n(x,\cdot) = g_n(x), \quad u_p(x,\cdot) = g_c(x), \quad u_{\varphi S}(x,\cdot) = g_S(x) + i_S(p_c)$$

at the contact areas Γ_{cS} of the substrate, the Dirichlet boundary conditions

$$u_{\varphi O}(x,\cdot) = g_O(x) + i_O(p_c)$$

at the contact areas Γ_{cO} of the substrate homogeneous Neumann boundary conditions for u_n, u_p and $u_{\varphi S}$ on the isolating boundary Γ_{NS} of the substrate, Neumann boundary conditions for $u_{\varphi S}$ on the isolating boundary Γ_{NO} and the boundary equations

$$f_{Jn}(u_n, u_{\varphi S}, x) \cdot \nu_S(x) = 0, \quad f_{Jp}(u_p, u_{\varphi S}, x) \cdot \nu_S(x) = 0.$$

at the interface Γ_I between oxide and substrate where $\nu_S(x)$ is the outer unit normal of the substrate at point x. The functions i_S and i_O are indicator functions selecting the contacts at the substrate and the oxide. The functions g_n, g_p, g_S and g_O are given functions of the space position x.

Cross Talking Lines Cross talking lines are multi-branch elements. Usually, they describe the mutual influence of two conduction branches and are therefore connected to four nodes of the network (see Fig. 5). If high frequency input signals are applied to circuits then we may observe a cross talk over tight lines. In order to model this cross talk we use the electromagnetic model developed in [3]. The model in [3] arises from the full Maxwell equations spatially discretized with the

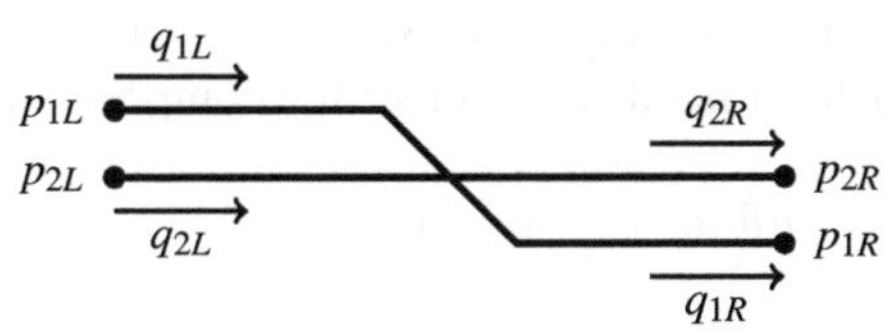

Fig. 5 Two cross talking branches have four contact nodes. Correspondingly, we have four pressure values p_{1L}, p_{1R}, p_{2L} and p_{2R} and four flows q_{1L}, q_{1R}, q_{2L} and q_{2R}

finite integration technique (FIT). The FIT discretization is an established tool to discretize electromagnetic devices which was developed and formulated by Thomas Weiland [10, 35–37].

$$
\begin{aligned}
(-q_{1L}, q_{1R}, -q_{2L}, q_{2R})^\top - \Lambda^\top C^\top M_\nu C u_A &= 0,\\
\vartheta \tilde{S} M_\epsilon G \frac{\mathrm{d}}{\mathrm{d}t} u_\phi + \tilde{S} M_{\bar{\nu}} u_A &= 0,\\
M_\epsilon \frac{\mathrm{d}}{\mathrm{d}t}(\frac{\mathrm{d}}{\mathrm{d}t} u_A + G u_\phi - \Lambda(p_{1L}, p_{1R}, p_{2L}, p_{2R})^\top) + C^\top M_\nu C u_A &\\
+ M_\sigma(\frac{\mathrm{d}}{\mathrm{d}t} u_A + G u_\phi - \Lambda(p_{1L}, p_{1R}, p_{2L}, p_{2R})^\top) &= 0.
\end{aligned}
$$

The discretized vector potential $u_A \in \mathbb{R}^{3n}$ and the discretized scalar potential $u_\phi \in \mathbb{R}^n$ are additional variables for the description of the flow/current relation. The dimension n depends on the refinement of the FIT discretization. The matrices $M_\epsilon, M_\sigma, M_\nu \in \mathbb{R}^{3n \times 3n}$ represent the three material properties for the permittivity, the conductivity and the reluctivity.

The matrix $G \in \mathbb{R}^{3n \times n}$ describes the discretized gradient operator, $\tilde{S} \in \mathbb{R}^{n \times 3n}$ is the discretized divergence operator and $C \in \mathbb{R}^{3n \times 3n}$ reflects discretized rotation operator.

The variable ϑ is used for different types of gauging. The case $\vartheta = 0$ describes the Coulomb gauge condition. The case $\vartheta = 1$ reflects the Lorenz gauge condition.

Finally, the excitation matrix $\Lambda \in \mathbb{R}^{3n \times 4}$ represents the boundary operator. It indicates which points of the discretization grid belong to one of the four contact areas. The transposed excitation matrix $\Lambda^\top \in \mathbb{R}^{4 \times 3n}$ represents a finite approximation of the integral over the four contact areas.

Gas Networks

Gas networks usually consist of pipes, valves, resistors, compressors, preheaters and coolers. We present here some model descriptions for the first three types of elements, while compressors, preheaters and coolers are often modeled by characteristic maps.

Pipe A pipe is a branch element. The flow through the gas pipe can be described on different model levels. As an example we use here the one-dimensional isothermal Euler equations [2, 5, 19] for the description of the compressible flow.

$$
\begin{aligned}
a\partial_t u_\varrho + \partial_x u_q &= 0\\
\partial_t u_q + \partial_x(u_p + \frac{u_q^2}{a u_\varrho}) &= -\frac{\lambda}{2D}\frac{u_q |u_q|}{a u_\varrho} - a g u_\varrho h'\\
u_p &= R u_\varrho T(1 + c_\alpha u_p)
\end{aligned}
\tag{24}
$$

with the pipe cross section area a, the pipe diameter D, the temperature T, the gas constant R, the gravity constant g, the pipe slope h' and the pipe friction coefficient λ and a real gas factor c_α. The pressure u_p, the gas density u_ϱ and the gas flow u_q through the pipe are time and position dependent. The network flows q and the node pressures p are related by the boundary conditions

$$\begin{aligned} u_p(x_L,t) &= p_L(t), & u_q(x_L,t) &= q_L(t), \\ u_p(x_R,t) &= p_R(t), & u_q(x_R,t) &= q_R(t). \end{aligned} \tag{25}$$

for all time points t.

Valve A valve is a branch element. It exists in different forms and can be modeled on different model levels. The easiest model describes a valve as a switch with two states *on* and *off*. The flow/pressure equations are then given by

$$\begin{cases} q = q_R = q_L, \quad p_L = p_R & \text{if the valve is } on \\ q = q_R = q_L = 0 & \text{if the valve is } off \end{cases} \tag{26}$$

Usually the state of the valve is controlled from outside and then given as a function of time. However, there are also self-controlled valves, for instance a non-return valve. It allows the flow to move into one direction only. We assume that the flow can move from left to right. Then,

$$\text{the non-return valve switches to } \begin{cases} on & \text{if } p_L > p_R \text{ and } q \geq 0 \\ off & \text{if } q < 0 \end{cases} \tag{27}$$

Resistor A resistor is a branch element. A resistor model is used to describe the hydraulic resistance of a valve. It specifies the pressure loss $\Delta p = p_L - p_R$ by

$$p_L - p_R = \frac{\zeta}{2} u_\varrho v^2 = \frac{\zeta}{2} \frac{q^2}{u_\varrho a^2}$$

with $q = q_L = q_R$, the pressure loss coefficient ζ, the velocity v, the gas density u_ϱ at the left node and the cross-section area a, see [30]. Assuming the gas compression factor to be constant, we have

$$p_L = u_\varrho c^2$$

with the constant sonic velocity c. It results in

$$p_L(p_L - p_R) = bq^2 \tag{28}$$

with the constant $b := \frac{c^2}{2a^2}\zeta$.

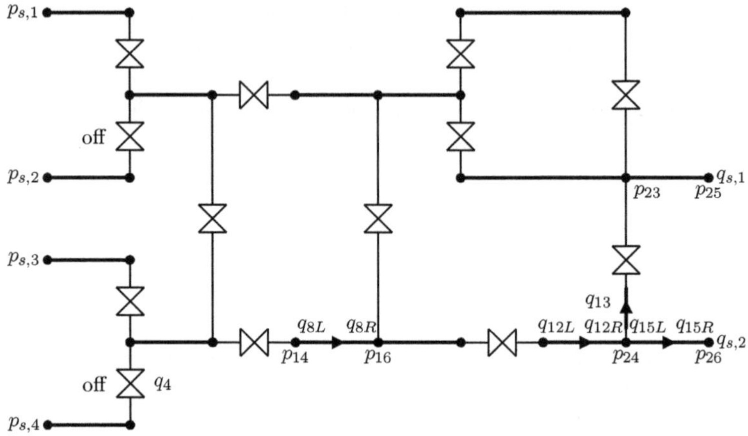

Fig. 6 Gas pipeline network with a gamma pulse characteristics. At the nodes 1–4 the pressure is fixed. The nodes 25 and 26 are sink nodes with a given flow demand. Two valves are closed. All other valves are open

A simple example for a gas pipe network with a gamma pulse characteristics is given in Fig. 6. It consists of 15 pipes and 13 valves. The network has four nodes at the left hand side where the pressure is fixed. At the right hand side we have two demand nodes with the demands $q_{s,1}$ and $q_{s,2}$. The pipe model equations (24) for the pipe between the nodes p_{14} and p_{16} are given by

$$a\partial_t u_{\varrho 8} + \partial_x u_{q8} = 0$$

$$\partial_t u_{q8} + \partial_x (u_{p8} + \frac{u_{q8}^2}{a u_{\varrho 8}}) = -\frac{\lambda}{2D}\frac{u_{q8}|u_{q8}|}{a u_{\varrho 8}} - agu_{\varrho 8}h'$$

$$u_{p8} = Ru_{\varrho 8}T(1 + c_\alpha u_{p8})$$

with the boundary conditions

$$u_{p8}(x_L, t) = p_{14}(t), \qquad u_{q8}(x_L, t) = q_{8L}(t),$$
$$u_{p8}(x_R, t) = p_{16}(t), \qquad u_{q8}(x_R, t) = q_{8R}(t).$$

Analogously, one can formulate the pipe equations for the other pipes in the network. For the open valve number 13 we get

$$q_{13} = q_{13R} = q_{13L}, \quad p_{23} = p_{24}.$$

Analogously, the equations of the other open valves can be formulated. For the lower closed valve we get

$$q_4 = q_{4R} = q_{4L} = 0$$

Analogously, the flow is zero through the other closed valve. The flow balance equation of the demand node 26 reads

$$q_{15R} = q_{s,2}.$$

Correspondingly, the flow balance equation for the other demand node 25 can be written. The flow balance equation of the node 24 is given by

$$q_{12R} - q_{13} - q_{15L} = 0.$$

Analogously, one can derive the flow balance equations of the remaining nodes.

Water Networks

Water supplying networks usually consist of pipes, valves, pumps, turbines, tanks and reservoirs. As an example we present models for pipes, pumps, tanks and reservoirs. Turbines are often modeled by characteristic maps. Valves are modeled in the same manner as for gas networks.

Pipe A pipe is a branch element. In the absence of shocks the incompressible flow can be described by the quasi-stationary model equations [21]

$$\frac{\mathrm{d}}{\mathrm{d}t}q + c_1(p_R - p_L) + c_2 q|q| = c_3 \tag{29}$$

with $q = q_L = q_R$ and c_1, c_2, c_3 being constants depending on the pipe diameter, length, cross-section area, friction and inclination angle.

Tank A tank is a node element.

$$a_R^\top q_R + a_L^\top q_L = \frac{\mathrm{d}}{\mathrm{d}t} f_c(p) \tag{30}$$

with a_R and a_L being the columns of A_L and A_R belonging to the node of the tank element. Correspondingly, q_R and q_L are the full vectors of left and right network flows, respectively. The function f_c describes the capacity of the tank.

Reservoir A reservoir is a node element. It has an unlimited capacity but constant pressure c_p. Correspondingly, we get the simple relation

$$p = c_p. \tag{31}$$

Pump A pump is a branch element. Its characteristic is usually described by an algebraic relation of the form

$$p_R - p_L = f_{pu}(q) \tag{32}$$

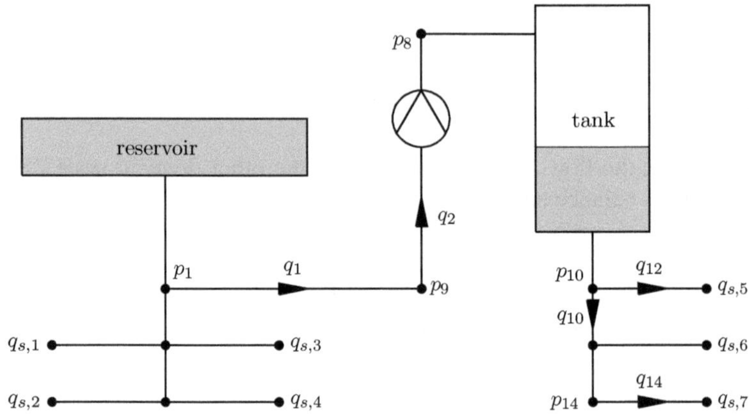

Fig. 7 Water pipeline network with one reservoir, one tank, one pump and 13 pipes. Seven nodes are demand nodes with a flow demand $q_{s,i}$ for $i = 1, \ldots, 7$

with $q = q_L = q_R$. One example is the representation in EPANET [15] as

$$p_R - p_L = aq^b$$

with a certain resistance coefficient a and flow exponent b.

One small example of a water network with a reservoir, one tank, one pump and 13 pipes is given in Fig. 7. The reservoir equation reads

$$p_1 = c_p$$

with the constant pressure c_p. Using the quasi-stationary model equations (29) for the pipes, we have

$$\frac{\mathrm{d}}{\mathrm{d}t} q_1 + c_1(p_1 - p_9) + c_2 q_1 |q_1| = c_3$$

for the pipe between the nodes p_1 and p_9. Analogously, one can describe the other 12 pipes. Taking the EPANET pump model [15], we get

$$p_8 - p_9 = aq_2^b$$

with a certain resistance coefficient a and a flow exponent b. The tank is a node element and may be described by

$$q_{10} + q_{12} = \frac{\mathrm{d}}{\mathrm{d}t} f_c(p_{10})$$

with the function f_c describing the capacity of the tank. The flow balance equation at the demand node with demand $q_{s,7}$ has the simple form

$$q_{14} = q_{s,7}.$$

Analogously, the flow balance equations for the other six demand nodes can be formulated. The flow balance equations for the rest of the nodes are just reflecting that the sum of all inflowing pipe flows equals the sum of all outflowing pipe flows.

2 Model Classes for Flow Networks

Depending on the different element models, the description of flow networks results in different types of equation systems: pure linear/nonlinear equation systems, differential algebraic equation systems, differential algebraic equation systems or partial differential algebraic equation systems.

Next, we want to elaborate some model classes reflecting topological net properties of flow networks.

2.1 *Static Networks with Lumped Element Models*

Lumped element models are characterized by constant flows on each network branch, i.e. $q_L = q_R = q$. Then, we have four types of net elements:

1. $$\Delta\tilde{p} = \tilde{p}_R - \tilde{p}_L = \tilde{f}_{pq}(\tilde{q}), \qquad (T_{pq})$$
2. $$\tilde{q} = \tilde{f}_{qp}(\Delta\tilde{p}) = \tilde{f}_{qp}(\tilde{p}_R - \tilde{p}_L), \qquad (T_{qp})$$
3. $$\Delta\tilde{p} = \tilde{p}_R - \tilde{p}_L = \tilde{c}_p, \qquad (T_p)$$
4. $$\tilde{q} = \tilde{c}_q \qquad (T_q)$$

with c_p and c_q being constants reflecting constant pressure and flow sources, respectively. The tilde notation is used to stress that we are describing a single element. The subscript in the types of elements shall indicate first the variables to be controlled and second the controlling variables. Summarizing the element equations (T_{pq}), (T_{qp}), (T_q), (T_p) and the mass flow balance equation (4), we obtain the system

$$A_{pq}^{\top} p = f_{pq}(q_{pq}) \qquad (33a)$$

$$q_{qp} = f_{qp}(A_{qp}^{\top} p) \qquad (33b)$$

$$A_p^\top p = c_p \tag{33c}$$

$$q_q = c_q \tag{33d}$$

$$A_{pq}q_{pq} + A_{qp}q_{qp} + A_p q_p + A_q q_q = 0 \tag{33e}$$

with the incidence matrices A_{pq}, A_{qp}, A_p and A_q collecting all columns for network elements of type (T_{pq}), (T_{qp}), (T_p) and (T_q), respectively. Obviously, they are related by

$$A = \begin{pmatrix} A_{pq} & A_{qp} & A_p & A_q \end{pmatrix}$$

Correspondingly, q_{pq}, q_{qp}, q_p and q_q denote the vectors of all flows of type (T_{pq}), (T_{qp}), (T_p) and (T_q), respectively.

For passive networks elements of type (T_{pq}) and (T_{qp}), the functions f_{pq} and f_{qp} are monotone.

Obviously, the system (33) can be reduced to the system

$$A_{qp} f_{qp}(A_{qp}^\top p) + A_{pq}q_{pq} + A_p q_p = -A_q c_q \tag{34a}$$

$$A_{pq}^\top p = f_{pq}(q_{pq}) \tag{34b}$$

$$A_p^\top p = c_p \tag{34c}$$

supplemented by the output equations

$$q_{qp} = f_{qp}(A_{qp}^\top p), \quad q_q = c_q \tag{35}$$

The equation system (34), (35) reflects resistive circuits with constant sources, static water supplying network models and static gas supplying network models.

2.2 Static Networks with Switching Element Models

We consider static networks with lumped element models supplemented with switching net elements of type

$$\tilde{s}(\tilde{p}_R - \tilde{p}_L, \tilde{q})(\tilde{p}_R - \tilde{p}_L) + (1 - \tilde{s}(\tilde{p}_R - \tilde{p}_L, \tilde{q}))\tilde{q} = 0, \tag{T_s}$$

with a switching function $\tilde{s}(\cdot,\cdot)$ having the values 1 or 0 only. For valves controlled from outside, the switching function $\tilde{s}$ is independent of $\Delta\tilde{p}$ and independent of $\tilde{q}$. Summarizing the element equations (T_{pq}), (T_{qp}), (T_q), (T_p), (T_s) and the mass flow balance equation (4), we obtain the system

$$A_{pq}^\top p = f_{pq}(q_{pq}) \tag{36a}$$

$$q_{qp} = f_{qp}(A_{qp}^\top p) \tag{36b}$$

$$A_p^\top p = c_p \tag{36c}$$

$$q_q = c_q \tag{36d}$$

$$S(A_s^\top p, q_s)A_s^\top p + (I - S(A_s^\top p, q_s))q_s = 0 \tag{36e}$$

$$A_{pq}q_{pq} + A_{qp}q_{qp} + A_p q_p + A_q q_q + A_s q_s = 0 \tag{36f}$$

with the incidence matrices A_{pq}, A_{qp}, A_p, A_q, A_s collecting all columns for network elements of type (T_{pq}), (T_{qp}) (T_p), (T_q), (T_s), respectively. Obviously, they are related by

$$A = \begin{pmatrix} A_{pq} & A_{qp} & A_p & A_q & A_s \end{pmatrix}$$

Correspondingly, q_{pq}, q_{qp}, q_p, q_q, q_s denote the vectors of all flows of type (T_{pq}), (T_{qp}), (T_p), (T_q) and (T_s), respectively. The matrix $S(A_s p, q_s)$ is a diagonal matrix with entries 1 or 0 only.

Obviously, the system (36) can be reduced to the system

$$A_{qp} f_{qp}(A_{qp}^\top p) + A_{pq}q_{pq} + A_p q_p + A_s(S(A_s^\top p, q_s)q_s) = -A_q c_q \tag{37a}$$

$$A_{pq}^\top p = f_{pq}(q_{pq}) \tag{37b}$$

$$A_p^\top p = c_p \tag{37c}$$

$$S(A_s^\top p, q_s)A_s^\top p = 0 \tag{37d}$$

supplemented by the output equations

$$q_{qp} = f_{qp}(A_{qp}^\top p), \quad q_q = c_q, \quad q_s = S(A_s^\top p, q_s)q_s \tag{38}$$

The equation system (37), (38) reflects static water supplying network models and static gas supplying network models with valves controlled from outside. The two latter ones are common models for the optimal control of such networks, see e.g. [7, 18].

2.3 *Dynamic Networks with Lumped Element Models*

As said before, lumped element models are characterized by constant flows on each network branch, i.e. $q_L = q_R = q$. Beside net elements of type (T_{qp}) and (T_{qp}), we have six additional types of elements:

1. $$\tilde{p}_R - \tilde{p}_L = \tilde{f}_{pt}(t), \tag{T_{pt}}$$
2. $$\tilde{q} = \tilde{f}_{qt}(t). \tag{T_{qt}}$$

3. $$\tilde{p}_R - \tilde{p}_L = \frac{\mathrm{d}}{\mathrm{d}t}\tilde{f}_{p\dot{q}}(\tilde{q}), \tag{$T_{p\dot{q}}$}$$

4. $$\tilde{q} = \frac{\mathrm{d}}{\mathrm{d}t}\tilde{f}_{q\dot{p}}(\tilde{p}_R - \tilde{p}_L), \tag{$T_{q\dot{p}}$}$$

5. $$\tilde{p}_R - \tilde{p}_L = \frac{\mathrm{d}}{\mathrm{d}t}\tilde{f}_{pq\dot{u}}(\tilde{u}), \quad \tilde{q} = \frac{\mathrm{d}}{\mathrm{d}t}\tilde{u}, \tag{$T_{pq\dot{u}}$}$$

6. $$\tilde{q} = \frac{\mathrm{d}}{\mathrm{d}t}\tilde{f}_{qp\dot{u}}(\tilde{u}), \quad \tilde{p}_R - \tilde{p}_L = \frac{\mathrm{d}}{\mathrm{d}t}\tilde{u}. \tag{$T_{qp\dot{u}}$}$$

The tilde notation is used to stress that we are describing a single element. Summarizing the element equations (T_{pq}), (T_{qp}), (T_{qt}), (T_{pt}), ($T_{p\dot{q}}$), ($T_{q\dot{p}}$), ($T_{pq\dot{u}}$), ($T_{qp\dot{u}}$), and the mass flow balance equation (4), we obtain the system

$$A_{pq}^{\top} p = f_{pq}(q_{pq}) \tag{39a}$$

$$q_{qp} = f_{qp}(A_{qp}^{\top} p) \tag{39b}$$

$$A_{pt}^{\top} p = f_{pt}(t) \tag{39c}$$

$$q_{qt} = f_{qt}(t) \tag{39d}$$

$$A_{p\dot{q}}^{\top} p = \frac{\mathrm{d}}{\mathrm{d}t} f_{p\dot{q}}(q_{dpq}) \tag{39e}$$

$$q_{q\dot{p}} = \frac{\mathrm{d}}{\mathrm{d}t} f_{q\dot{p}}(A_{q\dot{p}}^{\top} p) \tag{39f}$$

$$A_{pq\dot{u}}^{\top} p = \frac{\mathrm{d}}{\mathrm{d}t} f_{pq\dot{u}}(u_{pq\dot{u}}), \quad q_{pq\dot{u}} = \frac{\mathrm{d}}{\mathrm{d}t} u_{pq\dot{u}} \tag{39g}$$

$$q_{qp\dot{u}} = \frac{\mathrm{d}}{\mathrm{d}t} f_{qp\dot{u}}(u_{qp\dot{u}}), \quad A_{qp\dot{u}}^{\top} p = \frac{\mathrm{d}}{\mathrm{d}t} u_{qp\dot{u}} \tag{39h}$$

$$A_{pq}q_{pq} + A_{qp}q_{qp} + A_{pt}q_{pt} + A_{qt}q_{qt} \tag{39i}$$

$$+ A_{p\dot{q}}q_{p\dot{q}} + A_{q\dot{p}}q_{q\dot{p}} + A_{pq\dot{u}}q_{pq\dot{u}} + A_{qp\dot{u}}q_{qp\dot{u}} = 0 \tag{39j}$$

with the incidence matrices A_{pq}, A_{qp}, A_{pt}, A_{qt}, $A_{p\dot{q}}$, $A_{q\dot{p}}$, $A_{pq\dot{u}}$, $A_{qp\dot{u}}$ collecting all columns for network elements of type (T_{pq}), (T_{qp}) (T_{pt}), (T_{qt}), ($T_{p\dot{q}}$), ($T_{q\dot{p}}$) ($T_{pq\dot{u}}$) and ($T_{qp\dot{u}}$) respectively. Obviously, they are related by

$$A = \left(A_{pq}\ A_{qp}\ A_{pt}\ A_{qt}\ A_{p\dot{q}}\ A_{q\dot{p}}\ A_{pq\dot{u}}\ A_{qp\dot{u}}\right)$$

Correspondingly, q_{pq}, q_{qp}, q_{pt}, q_{qt}, $q_{p\dot{q}}$, $q_{q\dot{p}}$, $q_{pq\dot{u}}$ and $q_{qp\dot{u}}$ denote the vector of all flows of type (T_{pq}), (T_{qp}), (T_{pt}), (T_{qt}), ($T_{p\dot{q}}$), ($T_{q\dot{p}}$), ($T_{pq\dot{u}}$) and ($T_{qp\dot{u}}$) respectively.

Obviously, the system (39) can be reduced to the system

$$A_{q\dot{p}}\frac{\mathrm{d}}{\mathrm{d}t}f_{q\dot{p}}(A_{q\dot{p}}^{\top}p) + A_{pq\dot{u}}\frac{\mathrm{d}}{\mathrm{d}t}u_{pq\dot{u}} + A_{qp\dot{u}}\frac{\mathrm{d}}{\mathrm{d}t}f_{qp\dot{u}}(u_{qp\dot{u}})$$

$$+A_{qp}f_{qp}(A_{qp}^{\top}p) + A_{p\dot{q}}q_{p\dot{q}} + A_{pq}q_{pq} + A_{pt}q_{pt} = -A_{qt}f_{qt}(t), \quad (40a)$$

$$\frac{\mathrm{d}}{\mathrm{d}t}f_{pq\dot{u}}(u_{pq\dot{u}}) = A_{pq\dot{u}}^{\top}p, \quad \frac{\mathrm{d}}{\mathrm{d}t}u_{qp\dot{u}} = A_{qp\dot{u}}^{\top}p, \quad \frac{\mathrm{d}}{\mathrm{d}t}f_{p\dot{q}}(q_{p\dot{q}}) = A_{p\dot{q}}^{\top}p, \quad (40b)$$

$$A_{pq}^{\top}p = f_{pq}(q_{pq}), \quad A_{pt}^{\top}p = f_{pt}(t) \quad (40c)$$

supplemented by the output equations

$$q_{q\dot{p}} = \frac{\mathrm{d}}{\mathrm{d}t}f_{q\dot{p}}(A_{q\dot{p}}^{\top}p), \quad q_{pq\dot{u}} = \frac{\mathrm{d}}{\mathrm{d}t}u_{pq\dot{u}}, \quad q_{qp\dot{u}} = \frac{\mathrm{d}}{\mathrm{d}t}f_{qp\dot{u}}(u_{qp\dot{u}}), \quad (41a)$$

$$q_{qp} = f_{qp}(A_{qp}^{\top}p), \quad q_{qt} = f_{qt}(t) \quad (41b)$$

For passive networks elements of type (T_{pq}), (T_{qp}), ($T_{p\dot{q}}$), ($T_{q\dot{p}}$), ($T_{pq\dot{u}}$) and ($T_{qp\dot{u}}$), the functions f_{pq}, f_{qp} $f_{p\dot{q}}$, $f_{q\dot{p}}$, $f_{pq\dot{u}}$ and $f_{qp\dot{u}}$ are monotone.

In absence of elements of type ($T_{pq\dot{u}}$) and ($T_{pq\dot{u}}$) (for instance lumped circuit models without memristors and memductors), the system (40), (41) reduces to

$$A_{q\dot{p}}\frac{\mathrm{d}}{\mathrm{d}t}f_{q\dot{p}}(A_{q\dot{p}}^{\top}p) + A_{qp}f_{qp}(A_{qp}^{\top}p) + A_{p\dot{q}}q_{p\dot{q}} + A_{pq}q_{pq} + A_{pt}q_{pt} = -A_{qt}f_{qt}(t), \quad (42a)$$

$$\frac{\mathrm{d}}{\mathrm{d}t}f_{p\dot{q}}(q_{p\dot{q}}) = A_{p\dot{q}}^{\top}p, \quad A_{pq}^{\top}p = f_{pq}(q_{pq}), \quad A_{pt}^{\top}p = f_{pt}(t) \quad (42b)$$

supplemented by the output equations

$$q_{qp} = f_{qp}(A_{qp}^{\top}p), \quad q_{qt} = f_{qt}(t), \quad q_{q\dot{p}} = \frac{\mathrm{d}}{\mathrm{d}t}f_{q\dot{p}}(A_{q\dot{p}}^{\top}p). \quad (43a)$$

2.4 Dynamic Networks with Switching Element Models

We consider static networks with lumped element models supplemented with switching net elements of type (T_s) and type

$$\tilde{s}(t)(\tilde{p}_R - \tilde{p}_L) + (1 - \tilde{s}(t))\tilde{q} = 0, \quad (T_{st})$$

with a switching function $\tilde{s}(t)$ having the values 1 or 0 only. Elements of type (T_{st}) are valves controlled from outside, for instance controlled by a dispatcher. Similarly to the derived net equations in the sections before, we obtain the equation system

$$A_{q\dot{p}} \frac{\mathrm{d}}{\mathrm{d}t} f_{q\dot{p}}(A_{q\dot{p}}^{\top} p) + A_{pq\dot{u}} \frac{\mathrm{d}}{\mathrm{d}t} u_{pq\dot{u}} + A_{qp\dot{u}} \frac{\mathrm{d}}{\mathrm{d}t} f_{qp\dot{u}}(u_{qp\dot{u}}) + A_{qp} f_{qp}(A_{qp}^{\top} p) + A_{p\dot{q}} q_{p\dot{q}}$$

$$+ A_{pq} q_{pq} + A_{pt} q_{pt} + A_s(S(A_s^{\top} p, q_s) q_s) + A_{st}(S(t) q_{st}) = -A_{qt} f_{qt}(t), \tag{44a}$$

$$\frac{\mathrm{d}}{\mathrm{d}t} f_{pq\dot{u}}(u_{pq\dot{u}}) = A_{pq\dot{u}}^{\top} p, \quad \frac{\mathrm{d}}{\mathrm{d}t} u_{qp\dot{u}} = A_{qp\dot{u}}^{\top} p, \quad \frac{\mathrm{d}}{\mathrm{d}t} f_{p\dot{q}}(q_{p\dot{q}}) = A_{p\dot{q}}^{\top} p, \tag{44b}$$

$$A_{pq}^{\top} p = f_{pq}(q_{pq}), \;\; A_{pt}^{\top} p = f_{pt}(t), \;\; S(A_s^{\top} p, q_s) A_s^{\top} p = 0, \;\; S(t) A_{st}^{\top} p = 0 \tag{44c}$$

supplemented by the output equations

$$q_{q\dot{p}} = \frac{\mathrm{d}}{\mathrm{d}t} f_{q\dot{p}}(A_{q\dot{p}}^{\top} p), \quad q_{pq\dot{u}} = \frac{\mathrm{d}}{\mathrm{d}t} u_{pq\dot{u}}, \quad q_{qp\dot{u}} = \frac{\mathrm{d}}{\mathrm{d}t} f_{qp\dot{u}}(u_{qp\dot{u}}), \tag{45a}$$

$$q_{qp} = f_{qp}(A_{qp}^{\top} p), \quad q_{qt} = f_{qt}(t), \quad q_s = S(A_s^{\top} p, q_s) q_s, \quad q_{st} = S(t) q_{st}. \tag{45b}$$

The differential algebraic system (44) has the particular difficulty that the usual rank constancy required in the various index concepts for analyzing differential algebraic equations is not fulfilled anymore. Depending on the type of switching elements, the rank changes can be state or time dependent.

2.5 Dynamic Networks with Distributed Element Models

As we have seen in Sect. 1.2, some net elements are described by distributed models. Depending on the kind of model, they can be hyperbolic (e.g. supersonic flow models), parabolic or elliptic (e.g. subsonic flow models) or a mix of these types (e.g. drift-diffusion flow models). As an example we take here an abstraction from the one-dimensional gas pipe model description in section "Gas Networks". It results in elements of type

$$\begin{aligned} &\partial_t \tilde{u}_p + c \partial_x \tilde{u}_q = 0, \quad \partial_t \tilde{u}_q + \partial_x \tilde{f}(\tilde{u}_p, \tilde{u}_q) = \tilde{g}(\tilde{u}_p, \tilde{u}_q) \\ &\tilde{u}_p(x_L, \cdot) = \tilde{p}_L, \;\; \tilde{u}_p(x_R, \cdot) = \tilde{p}_R, \;\; \tilde{u}_q(x_L, \cdot) = \tilde{q}_L, \;\; \tilde{u}_q(x_R, \cdot) = \tilde{q}_R. \end{aligned} \tag{$T_{\partial_t \partial_x}$}$$

with a function $f(u_p, u_q)$ being positive for $u_p > 0$ and a function $g(u_p, u_q)$ that is decreasing with respect to u_q and increasing with respect to u_p. Collecting them for all network branches and coupling them with the mass balance equation yields the system

$$\partial_t u_p + C \partial_x u_q = 0, \quad \partial_t u_q + \partial_x f(u_p, u_q) = g(u_p, u_q), \quad A_L q_L + A_R q_R = 0, \tag{46a}$$

$$u_p(x_L, \cdot) = -A_L^{\top} p, \;\; u_p(x_R, \cdot) = A_R^{\top} p, \;\; u_q(x_L, \cdot) = q_L, \;\; u_q(x_R, \cdot) = q_R. \tag{46b}$$

Here u_p and u_q denote the vectors of the pressure and the flow of all branches. The evaluation at point x_L/x_R means the evaluation for each branch at their left/right node. C is a constant diagonal matrix. The system (46a)–(46b) can be reduced to the system

$$\partial_t u_p + C\,\partial_x u_q = 0, \quad \partial_t u_q + \partial_x f(u_p, u_q) = g(u_p, u_q), \tag{47a}$$

$$A_L u_q(x_L, \cdot) + A_R u_q(x_R, \cdot) = 0, \;\; u_p(x_L, \cdot) = -A_L^\top p, \;\; u_p(x_R, \cdot) = A_R^\top p \tag{47b}$$

with the output equations

$$q_L = u_q(x_L, \cdot), \quad q_R = u_q(x_R, \cdot). \tag{48}$$

Conclusions

Flow networks can be described by network graphs with node variables p reflecting the pressure at each node and edge variables q reflecting the flow at each edge. Depending on the kind of net element models, the resulting systems are linear/nonlinear equations, differential algebraic equations or partial differential algebraic equations.

Typical kinds of network models (static/dynamic with lumped/switching/distributed/discretized net element models) have been presented with its inner structure. The common framework described here allows a rigorous network analysis independently of the type of flow media but in dependence of the kind of net element models.

Additionally, it allows the transfer of numerical and optimization methods developed for one kind of network to another kind of network (e.g. circuit methods to methods for water/gas networks) as long as they contain the same type of net element models.

Finally, we particularly want to note that the analytical treatment, numerical simulation and optimal control of network models with switched or distributed net element descriptions requires new developments in the numerical analysis and optimal control of switched differential algebraic systems or partial differential algebraic systems with the structures presented here.

References

1. Ahmed, M., Cho, K., Cho, T.W.: Memristance and memcapacitance modeling of thin film devices showing memristive behavior. In: 2012 13th International Workshop on Cellular Nanoscale Networks and Their Applications (CNNA), Turin, pp. 1–5 (2012). doi:10.1109/CNNA.2012.6331436

2. Bales, P., Kolb, O., Lang, J.: Hierarchical modelling and model adaptivity for gas flow on networks. In: Allen, G., Nabrzyski, J., Seidel, E., Albada, G.D., Dongarra, J., Sloot, P.M. (eds.) Computational Science – ICCS 2009. 9th International Conference Baton Rouge, LA, USA, May 25–27, 2009. Lecture Notes in Computer Science, vol. 5544, pp. 337–346. Springer, Berlin/Heidelberg (2009). doi:10.1007/978-3-642-01970-8_33
3. Baumanns, S.: Coupled electromagnetic field/circuit simulation: modeling and numerical analysis. Ph.D. thesis, University of Cologne (2012)
4. Baumanns, S., Jansen, L., Selva Soto, M., Tischendorf, C.: Analysis of semi-discretized differential algebraic equation from coupled circuit device simulation. Computational and Applied Mathematics, Springer Basel, pp. 1–23, (2014). doi:10.1007/s40314-014-0157-4
5. Brouwer, J., Gasser, I., Herty, M.: Gas pipeline models revisited: model hierarchies, nonisothermal models, and simulations of networks. Multiscale Model. Simul. **9**(2), 601–623 (2011). doi:10.1137/100813580
6. BSIM Group: Providing the world with transistor models for IC design. (2014). http://www-device.eecs.berkeley.edu/bsim/
7. Burgschweiger, J., Gnädig, B., Steinbach, M.: Optimization models for operative planning in drinking water networks. Optim. Eng. **10**(1), 43–73 (2009). doi:10.1007/s11081-008-9040-8
8. Chua, L.: Resistance switching memories are memristors. Appl. Phys. A: Mater. Sci. Process. **102**(4), 765–783 (2011)
9. Chua, L., Desoer, C., Kuh, E.: Linear and Nonlinear Circuits. McGraw-Hill Book, Singapore (1987)
10. Clemens, M., Weiland, T.: Discrete electromagnetism with the finite integration technique. Prog. Electromagn. Res. (PIER) **32**, 65–87 (2001)
11. Danielsen, M., Ottesen, J.T.: 6. A cardiovascular model, chap. 6. In: Ottesen, J.T., Olufsen, M.S., Larsen, J. (eds.) Applied Mathematical Models in Human Physiology, pp. 137–155. Society for Industrial and Applied Mathematics, Philadelphia (2004). doi:10.1137/1.9780898718287.ch6
12. Desoer, C., Kuh, E.: Basic Circuit Theory. International student edition. McGraw-Hill, Auckland/Singapore (1984)
13. Deuerlein, J.: Decomposition model of a general water supply network graph. J. Hydraul. Eng. **134**(6), 822–832 (2008). doi:10.1061/(ASCE)0733-9429(2008)134:6(822)
14. Di Ventra, M., Pershin, Y., Chua, L.: Circuit elements with memory: memristors, memcapacitors, and meminductors. Proc. IEEE **97**(10), 1717–1724 (2009). doi:10.1109/JPROC.2009.2021077
15. EPANET: Software that models the hydraulic and water quality behavior of water distribution piping systems. (2014). http://www.epa.gov/nrmrl/wswrd/dw/epanet.html
16. Estévez Schwarz, D., Tischendorf, C.: Structural analysis of electric circuits and consequences for MNA. Int. J. Circuit Theory Appl. **28**(2), 131–162 (2000)
17. Gajewski, H., Gröger, K.: On the basic equations for carrier transport in semiconductors. J. Math. Anal. Appl. **113**, 12–35 (1986)
18. Gugat, M., Leugering, G., Schittkowski, K., Schmidt, E.: Modelling, stabilization, and control of flow in networks of open channels. In: Grötschel, M., Krumke, S., Rambau, J. (eds.) Online Optimization of Large Scale Systems, pp. 251–270. Springer, Berlin/Heidelberg (2001). doi:10.1007/978-3-662-04331-8_16
19. Herty, M., Mohring, J., Sachers, V.: A new model for gas flow in pipe networks. Math. Methods Appl. Sci. **33**(7), 845–855 (2010). doi:10.1002/mma.1197. http://dx.doi.org/10.1002/mma.1197
20. Jansen, L., Matthes, M., Tischendorf, C.: Global unique solvability for memristive circuit DAEs of index 1. Int. J. Circuit Theory Appl., (2013). doi:10.1002/cta.1927
21. Jansen, L., Pade, J.: Global unique solvability for a quasi-stationary water network model. Preprint 2013–11, Dept. of Math., Humboldt-Universität zu Berlin, (2013)
22. Liu, W.: MOSFET Models for SPICE Simulation: Including BSIM3v3 and BSIM4. Wiley-IEEE, New York (2001)

23. Markowich, P.A., Ringhofer, C.A., Schmeiser, C.: Semiconductor Equations. Springer, Wien (1990)
24. Quarteroni, A., Ragni, S., Veneziani, A.: Coupling between lumped and distributed models for blood flow problems. Comput. Vis. Sci. **4**(2), 111–124 (2001)
25. Riaza, R.: Dynamical properties of electrical circuits with fully nonlinear memristors. Nonlinear Anal.: Real World Appl. **12**(6), 3674–3686 (2011)
26. Riaza, R.: Manifolds of equilibria and bifurcations without parameters in memristive circuits. SIAM J. Appl. Math. **72**(3), 877–896 (2012). doi:10.1137/100816559
27. Riaza, R.: First order mem-circuits: modeling, nonlinear oscillations and bifurcations. IEEE Trans. Circuits Systems I: Regul. Pap. **60**(6), 1570–1583 (2013). doi:10.1109/TCSI.2012.2221174
28. Riaza, R., Tischendorf, C.: Semistate models of electrical circuits including memristors. Int. J. Circuit Theory Appl. **39**(6), 607–627 (2011)
29. Selva Soto, M., Tischendorf, C.: Numerical analysis of DAEs from coupled circuit and semiconductor simulation. Appl. Numer. Math. **53**(2–), 471–488 (2005)
30. SIMONE: SIMONE Research Group S.R.O. (2014). www.simone.eu
31. Simpson, A.R.: Comparing the Q-equations and Todini-Pilati formulation for solving the water distribution system equations, chap. 5. In: Lansey, K.E., Choi, C.Y., Ostfeld, A., Pepper, I.L. (eds.) Proceedings of the 12th Annual Conference on Water Distribution Systems Analysis 2010, Tucson, pp. 37–54 (2010). doi:10.1061/41203(425)6
32. Tischendorf, C.: Modeling circuit systems coupled with distributed semiconductor equations. In: Antreich, K., Bulirsch, R., Gilg, A., Rentrop, P. (eds.) Mathematical Modeling, Simulation and Optimization of Integrated Electrical Circuits. No. 146 in International Series of Numerical Mathematics, pp. 229–247. Birkhäuser, Basel (2003)
33. Tischendorf, C.: Coupled systems of differential-algebraic and partial differential equations in circuit and device simulation. Modeling and numerical analysis (2004). Habilitation thesis at Humboldt University of Berlin
34. Todini, E., Pilati, S.: A gradient algorithm for the analysis of pipe networks. In: Coulbeck, B., Orr, C.-H. (eds.) Computer Applications in Water Supply: Vol. 1—Systems Analysis and Simulation, pp. 1–20. Research Studies Press, Taunton, (1988)
35. Weiland, T.: A discretization model for the solution of Maxwell's equations for six-component fields. Arch. Elektron. Übertrag. **31**(3), 116–120 (1977)
36. Weiland, T.: Time domain electromagnetic field computation with finite difference methods. Int. J. Numer. Model.: Electron. Netw. Devices Fields **9**(4), 295–319 (1996)
37. Yee, K.: Numerical solution of initial boundary value problems involving Maxwell's equations in isotropic media. IEEE Trans. Antennas Propag. **14**(3), 302–307 (1966)

Part III
Model Order Reduction

Index-Aware Model Order Reduction for Higher Index DAEs

Nicodemus Banagaaya and Wil H.A. Schilders

Abstract There exists many Model Order Reduction (MOR) methods for ODEs but little had been done to reduce DAEs especially higher index DAEs. In principle, if the matrix pencil of a DAE is regular, it is possible to use conventional MOR techniques to obtain reduced order models, which are generally ODEs. However, as far as their numerical treatment is concerned, the reduced models may be close to higher index models, that is, to DAEs. Thus the numerical solution of the reduced models might be computationally expensive, or even not feasible. In the worst cases, the reduced models may be unsolvable, i.e. their matrix pencil is singular. This problem is very pronounced for systems with index higher than 1, but it may occur even if the index of the problem does not exceed 1. Thus MOR methods for ODEs cannot generally be used for DAEs. This motivated us to introduce a new MOR method for DAEs which we call the index-aware MOR (IMOR) which can reduce DAEs while preserving the index of the system. This method involves first splitting the DAEs into differential and algebraic parts. Then, we use the existing MOR methods to reduce the differential part. We observed that the reduction of the differential part induces a reduction in the algebraic part. This enabled us to construct a method which reduces both the differential and the algebraic part. As a result a DAE is reduced. This method can also be used as a new method to solve DAEs. In this paper, we generalize the IMOR method to higher index DAEs and we shall call this method the GIMOR method. We use index-3 systems for testing and validating the accuracy of the GIMOR method.

Keywords Model order reduction • Tractability index • Special projectors • Special bases

Mathematics Subject Classification (2010) 78M34 · 65L80

N. Banagaaya (✉) • W.H.A. Schilders
Department of Mathematics and Computer Science, Eindhoven university of Technology, Eindhoven, The Netherlands
e-mail: n.banagaaya@tue.nl; w.h.a.schilders@tue.nl

S. Schöps et al. (eds.), *Progress in Differential-Algebraic Equations*, Differential-Algebraic Equations Forum, DOI 10.1007/978-3-662-44926-4_8

1 Introduction

Consider a control problem

$$Ex' = Ax + Bu, \quad x(0) = x_0, \tag{1a}$$

$$y = C^T x, \tag{1b}$$

where matrices $E, A \in \mathbb{R}^{n,n}$, $B \in \mathbb{R}^{n,m}$ and $C \in \mathbb{R}^{n,\ell}$ is the control output matrix. $x \in \mathbb{R}^n$ is the state space vector, $u \in \mathbb{R}^m$ is the input data vector, $y \in \mathbb{R}^\ell$ is the output data or desired solutions and $x_0 \in \mathbb{R}^n$ is the initial value. If E is non-singular, then (1a) is a system of ordinary differential equations (ODEs) otherwise it is a system of differential algebraic equations (DAEs). In this paper, we assume that E is singular, thus all our systems are DAEs unless otherwise stated. The corresponding $(E, \ A)$ is called the matrix pair or matrix pencil associated with the system and sometime written as $\lambda E - A$, where $\lambda \in \mathbb{C}$. The spectrum of the matrix pencil $(E, \ A)$ is defined by $\sigma(E, A) := \sigma_f(E, A) \cup \{\infty\}$, where $\sigma_f(E, A)$ are the finite eigenvalues of the matrix pair $(E, \ A)$ defined as $\sigma_f(E, A) = \{\lambda | \lambda \in \mathbb{C}, \quad \lambda \text{ finite}, \quad \det(\lambda E - A) = 0\}$. In order to ensure existence and uniqueness of the solutions of (1a), we assume that it's matrix pencil is regular, that is its polynomial $\mathscr{P}(\lambda) = \det(\lambda E - A)$ does not vanish identically. Moreover, DAEs are known to be very difficult to solve numerically due to the sensitivity of their solutions to perturbations. This sensitivity is measured in terms of the *index concepts*. In this paper, we use the *tractability index* because of being numerically feasible, but there are many other index concepts such as differentiation index, Kronecker index, strangeness index and so on. We note that all index concepts coincide in case of linear DAEs with constant coefficient matrices. Additional requirements are that the initial value x_0 must be a consistent initial condition and input data u must be smooth enough, i.e. if μ is the index of the DAE, then u must be at least $\mu - 1$ times differentiable. Systems of the form (1) arise in many fields such as power systems, electrical networks, aerospace engineering, constrained mechanical systems and so on. These systems tend to be very large and simulations of their solutions in real time is impossible due to very high storage and computational costs.

In practice n is very large compared to the desired output solutions ℓ, these kind of problems tend to be expensive to solve in real time. This is an attractive feature to apply model order reduction (MOR). In most recent years much research has been done in the area of model order reduction, consequently a large variety of MOR methods are available, some are tailored to specific application and other are more general [11]. All these methods can be categorized into two basic types, which are the balanced truncation and Krylov subspace based methods. The former is more accurate than the latter although it is computationally very expensive [11]. However, all these methods focus on the reduction of ordinary differential equations (ODEs) and little has been done to reduce DAEs. Krylov subspace based methods such as the PRIMA method [8] have been widely used to reduce DAEs. Unfortunately PRIMA method cannot always be used because the derived reduced-order models may lead to wrong solutions or very difficult to solve as discussed in [1, 2], especially for DAEs with index greater than 1. However, in [2], we discussed that the conventional methods can be applied even on the higher index DAEs depending on the nature of the consistent initial condition. For example if the consistent initial condition do not depend on the derivatives of the input data. The most recent attempts to reduce DAEs can be found in [3, 7, 9]. All these attempts are based on balanced truncation method and they are problem specific. This motivated us to propose a new MOR method specifically for DAEs which we call the index-aware MOR (IMOR) method. This method was first proposed in [1] and [2] for index-1 and 2 systems, respectively. The IMOR method involves first splitting the DAE into differential and algebraic parts, then we apply the convention MOR methods for ODEs on the differential part and then also reduce the algebraic part. In [1, 2], we observed that the reduction in the differential part induces a reduction in the algebraic part. Thus the reduction of the algebraic part depend on the reduction in the differential part. The splitting is based on the decoupling procedure proposed by März [6] using matrix and projector chain. However this decoupling procedure leads to much larger systems of dimension $n(\mu + 1)$ which is impractical to apply MOR. Thus, in [1, 2], we modified the März decoupling procedure using the special basis chains of the projectors. In this way we were able to preserve the dimension of the DAE as well as the spectrum of its matrix pencil.

In this paper, we generalize the IMOR method leading to the method which we call the generalized index-aware (GIMOR) method. Although, we illustrate it using the decoupling procedure on index-3 system but the same procedure can be extended to even higher index systems. We derived the GIMOR method based on the ideas proposed in [1, 2] for index-1 and 2 systems. This paper is arranged as follows: In Sect. 2, we briefly discuss the limitation of convention MOR method focusing on the PRIMA method [8]. Then, in Sect. 3, we discuss the decoupling of index-3 systems using special basis chains of projectors. We shall then propose the GIMOR method in Sect. 4.1. Finally, we apply the GIMOR method on index-3 systems.

2 Limitation of Conventional MOR Methods

In this section, we briefly introduce and discuss the limitation of conventional MOR methods on higher index DAEs. The main goal of MOR is to find a smaller dimension $r < n$ system of (1) which is computationally cheaper while preserving both the physical and mathematical properties of the original system. In our discussion, we shall restrict ourselves on the Krylov subspace based methods and our method of choice will be the PRIMA method [8]. Thus using the Arnold process to construct an orthonormal basis matrix $V_r \in \mathbb{R}^{n,r}$ such that $V_r^T V_r = \mathrm{I}_r$.

We can approximate the state vector $x = V_r x_r$. Substituting $x = V_r x_r$ into (1) and simplifying, we obtain a reduced-order model given by

$$E_r x_r' = A_r x_r + B_r u, \quad x_r(0) = x_{r_0}, \tag{2a}$$

$$y_r = C_r^T x_r, \tag{2b}$$

where $E_r = V_r^T E V_r$, $A_r = V_r^T A V_r \in \mathbb{R}^{r,r}$, $B_r = V_r^T B \in \mathbb{R}^{r,m}$, and $C_r = V_r^T C \in \mathbb{R}^{r,\ell}$. The accuracy of the reduced-order model (2) is measured by comparing its transfer function with that of the original model (1). The transfer function of (1) is obtained by taking it's Laplace transform which leads to $Y(s) = H(s)U(s) + Y_0(s)$, where $H(s) = C^T(sE - A)^{-1}B$ and $Y_0(s) = C^T(sE - A)^{-1}Ex(0)$. Then, we assume vanishing initial condition, i.e. $Y_0(s) = 0$ to obtain $Y(s) = H(s)U(s)$ and $H(s)$ is what is traditionally called the transfer function. Then an accurate reduced-order model is the one with a very small approximate error $\|H - H_r\|$ and $\|y - y_r\|$ in a suitable norm. However if E is singular, we cannot always have enough freedom to choose the initial condition especially for higher index DAEs. Hence applying PRIMA method directly on DAEs can lead to wrong or unsolvable reduced-order models as illustrated in the example below.

Example 1 In this example, we use the generator model originating from [4], as described in Fig. 1. In this model: The input function is the angle ϕ_{in} on the left axis. This axis is connected to a rotating mass with inertia J which is rotated at an angle ϕ and rotates with the angular velocity ω. The torque acting on the left side of the mass is M_1 and the torque on the right side is M_2. The mass is then connected to a second axis which is connected to the actual generator. The variables describing the second axis and the electrical quantities are then assumed to depend on each other according to $M_2 = k\mathrm{I}$ and $u_1 = k\omega$ for some constant k. The rest of the electrical circuit consists of two resistors and one inductor. The measured output

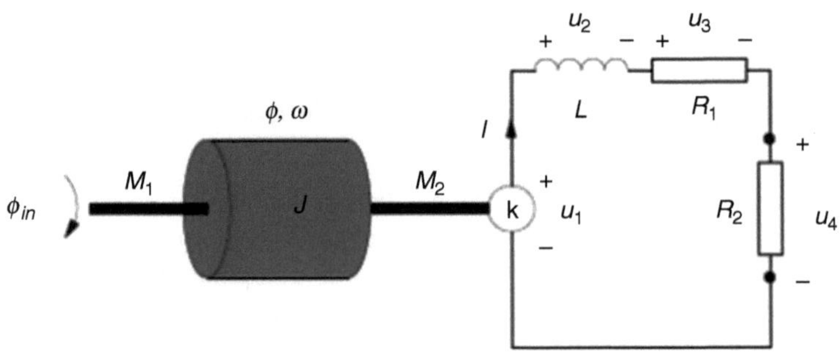

Fig. 1 A model of a generator

is the voltage u_4. This model leads to a DAE system of the form (1) with system matrices:

$$E = \begin{bmatrix} 0&0&0&0&0&0&0&0&0 \\ 0&0&0&1&0&0&0&0&0 \\ 0&0&J&0&0&0&0&0&0 \\ 0&0&0&0&0&0&0&0&0 \\ 0&0&0&0&0&0&0&0&0 \\ 0&0&0&0&L&0&0&0&0 \\ 0&0&0&0&0&0&0&0&0 \\ 0&0&0&0&0&0&0&0&0 \\ 0&0&0&0&0&0&0&0&0 \end{bmatrix}, \quad A = \begin{bmatrix} 0&0&0&-1&0&0&0&0&0 \\ 0&0&1&0&0&0&0&0&0 \\ 1&1&0&0&0&0&0&0&0 \\ 0&-1&0&0&k&0&0&0&0 \\ 0&0&k&0&0&-1&0&0&0 \\ 0&0&0&0&0&0&1&0&0 \\ 0&0&0&0&R_1&0&0&-1&0 \\ 0&0&0&0&R_2&0&0&0&-1 \\ 0&0&0&0&0&-1&1&1&1 \end{bmatrix},$$

$$B = \begin{bmatrix} 1\,0\,0\,0\,0\,0\,0\,0\,0 \end{bmatrix}^T, \quad C = \begin{bmatrix} 0\,0\,0\,0\,0\,0\,0\,0\,1 \end{bmatrix}^T,$$

$$x = \begin{bmatrix} M_1\; M_2\; \omega\; \phi\; \mathrm{I}\; u_1\; u_2\; u_3\; u_4 \end{bmatrix}^T \quad \text{and} \quad u = \phi_{in}. \tag{3}$$

This system is of dimension $n = 9$ and inputs $m = 1$ and outputs $\ell = 1$. This system is solvable since $\det(\lambda E - A) = -R_1 - R_2 - \lambda L \neq 0 \;\forall\; \lambda \in \mathbb{C}$. This implies that the finite spectrum of matrix pencil (E, A) is given by $\sigma_f(E, A) = \{-\frac{R_1+R_2}{L}\}$. Thus the DAE system is stable. Letting $J = 1, k = -1, R_1 = 1, R_2 = 1, L = 1$ and using $s_0 = 0$ as the expansion point, we were able construct the orthonormal basis matrix V_r using the PRIMA method.We then used this V_r to construct PRIMA reduced-order model of (3) given by:

$$E_r = \begin{bmatrix} 0 & 0 & 0 & 0 \\ -0.2774 & 0.4615 & -0.1155 & 0.0665 \\ -0.0595 & -0.2227 & 0.0557 & -0.0321 \\ -0.2637 & -0.1197 & 0.0299 & -0.0172 \end{bmatrix},$$

$$A_r = \begin{bmatrix} 0 & 0 & 0 & 0 \\ -0.2774 & 0.1538 & -0.7175 & 0 \\ -0.8326 & 0.0330 & 0.2944 & -0.0278 \\ 0.4795 & 0.1463 & 0.0962 & -0.0483 \end{bmatrix}, \tag{4}$$

$$B_r = \begin{bmatrix} 0\; 0.2774\; 0.8326\; -0.4795 \end{bmatrix}^T, \quad \text{and}$$

$$C_r = \begin{bmatrix} 0\; 0.2774\; 0.0595\; 0.2637 \end{bmatrix}^T.$$

We can see that the original model is reduced to dimension 4. The next step is to check the validity of the derived reduced-order model. Unfortunately the reduced model is unsolvable since $\det(\lambda E_r - A_r) = 0$. Thus we cannot use the PRIMA method to reduce system (3).

3 Decoupling of Index-3 DAEs

In this section, we discuss the decoupling of index-3 system using special bases of projectors as introduced in [1, 2]. The basic idea is based on the splitting technique introduced by März [6] using canonical projectors. Assume (1a) is of tractability index μ, then we can construct matrix and projector chain as follows: Set $E_0 := E,\ A_0 := A$, then iterate:

$$E_{j+1} = E_j - A_j Q_j, \quad A_{j+1} = A_j P_j, \quad j \geq 0, \tag{5}$$

till it converges to a non-singular matrix, i.e. $E_{\mu+j} = E_\mu$, where Q_j is a non unique projector chosen such that im Q_j = ker E_j and its complementary projector $P_j = \mathrm{I} - Q_j$. Thus μ is what we call the tractability index of the DAE system (1a). If $\mu = 1$, we call system (1a) an index-1 DAE system and if $\mu > 1$ we call it a higher index DAE system. According [6], using the final matrices $E_\mu, A_\mu \in \mathbb{R}^{n,n}$ and the projector chains $Q_j, P_j \in \mathbb{R}^{n,n}$, $j = 0, 1, \cdots, \mu - 1$, the DAE system (1a) can be written as

$$P_{\mu-1} \cdots P_0 x' + Q_0 x + \cdots + Q_{\mu-1} x = E_\mu^{-1} \left[A_\mu x + Bu \right], \tag{6}$$

if the additional constraints

$$Q_j Q_i = 0, \quad j > i, \tag{7}$$

hold. Using März decoupling procedure [6], system (6) can be decoupled in to 1 differential part and μ algebraic parts. However this decoupling procedure leads to a much larger system of dimension $n(\mu+1)$ which increases as the index μ increases. This motivated us to use bases of the projectors instead of the full projectors as proposed in [1, 2].

In our discussion, we shall focus on the case of index-3 systems, i.e. $\mu = 3$ but the same approach can be applied to even higher index DAEs, thus (6) can be written as

$$P_2 P_1 P_0 x' + Q_0 x + Q_1 x + Q_2 x = E_3^{-1} \left[A_3 x + Bu \right]. \tag{8}$$

If we use the März decoupling procedure this will lead to a decoupled system of dimension $4n$. The März decoupling procedure for index-3 systems can be modified as follows: Assume (7) holds, then the identity matrix can written into two decompositions given by

$$\mathrm{I} = P_0 + Q_0 = P_0 P_1 P_2 + P_0 P_1 Q_2 + P_0 Q_1 + Q_0, \tag{9a}$$

$$\mathrm{I} = P_2 + Q_2 = P_0 P_1 P_2 + Q_0 P_1 P_2 + Q_1 P_2 + Q_2. \tag{9b}$$

The first decomposition (9a) is used to define the differential and algebraic components:

$$x_P := P_0P_1P_2x, \quad x_{Q,0} := Q_0x, \quad x_{Q,1} := P_0Q_1x, \quad x_{Q,2} := P_0P_1Q_2x. \tag{10}$$

while the second decomposition (9b) is used to derive the differential and algebraic equations of the projected system (8). From (10), we observe that every variable is of dimension n, which means that the total dimension of the decoupled system will be $4n$. In order to modify this decoupling procedure, we use (9a) to construct the bases of projectors $\{Q_0, P_0Q_1, P_0P_1Q_2, P_0P_1P_2\}$ which are used to decouple (8) instead of the full projectors. In [2], we mentioned that higher index DAEs have a possibility of having a purely algebraic system depending to the nature of the spectrum of the matrix pencil $(E,\ A)$. Thus, if $\sigma_f(E,A) = \emptyset$, i.e. $\det(\lambda E - A) = c \in \mathbb{C}\setminus\{0\}$, $\forall\lambda \in \mathbb{C}$. This implies the projector product $P_0P_1P_2$ vanishes to zero which contributes to the differential part of the system. Thus in this section, we consider two cases of decomposition of the DAE system (1a) depending on the spectrum of the matrix pencil $(E,\ A)$. In both cases the starting point is the same as that of index 1 as presented in [1].

3.1 Index-3 Systems with a Differential Part

In this section, we assume that $P_0P_1P_2 \neq 0$. Thus, we need to construct the bases of projectors $\{Q_0, P_0Q_1, P_0P_1Q_2, P_0P_1P_2\}$. This done as follows: Let $k_0 = \dim(\ker E_0)$, $n_0 = n - k_0$, and consider a basis matrix $(p_0, q_0) = (p_{0,1}, \ldots, p_{0,n_0}, q_{0,1}, \ldots, q_{0,k_0}) \in \mathbb{R}^n$ which contains k_0 independent vectors $q_{0,i}$ which span $\ker E_0$. Since (p_0, q_0) is a basis matrix, it is non-singular, then we can let $(p_0, q_0)^{-1} = (p_0^*, q_0^*)^T$ be its inverse, where $q_0^* \in \mathbb{R}^{n,k_0}$ and $p_0^* \in \mathbb{R}^{n,n_0}$. Then, we have

$$q_0^{*T}q_0 = I_{k_0}, \quad q_0^{*T}p_0 = 0, \quad p_0^{*T}q_0 = 0, \quad p_0^{*T}p_0 = I_{n_0}. \tag{11}$$

We note that, by construction, we have

$$Q_0q_0 = q_0, \quad Q_0p_0 = 0, \quad P_0q_0 = 0, \quad P_0p_0 = p_0. \tag{12}$$

If we left and right multiply (9a) by p_0^{*T} and p_0, we obtain

$$I_{n_0} = Z_{q_0} + Z_{p_0}, \tag{13}$$

where $Z_{p_0} := p_0^{*T}P_0P_1Q_2p_0 + p_0^{*T}P_0P_1P_2p_0 \in \mathbb{R}^{n_0,n_0}$ and $Z_{q_0} := p_0^{*T}P_0Q_1p_0 \in \mathbb{R}^{n_0,n_0}$. Using the general form of Proposition 2.1 in [2], Z_{p_0} and Z_{q_0} are mutually orthogonal projectors, acting in $\mathbb{R}^{n_0}$. Next, we construct bases for projectors Z_{p_0} and Z_{q_0}. If, we let $k_1 = \dim(\operatorname{im} Z_{q_0})$, and $n_1 = n_0 - k_1$, and consider a basis matrix $(z_{p_0}, z_{q_0}) \in \mathbb{R}^{n_0}$ made of n_1 independent columns of projection matrix Z_{p_0} and k_1 independent columns of the complementary projection matrix Z_{q_0}. We denote by $(z^*_{p_0}, z^*_{q_0})^T$ the inverse of (z_{p_0}, z_{q_0}), such that

$$z^{*T}_{p_0} z_{p_0} = I_{n_1}, \quad z^{*T}_{p_0} z_{q_0} = 0, \quad z^{*T}_{q_0} z_{p_0} = 0, \quad z^{*T}_{q_0} z_{q_0} = I_{k_1}. \tag{14}$$

Using the identities (14) and (11) on (13) leads to

$$I_{n_1} = Z_{q_1} + Z_{p_1}, \tag{15}$$

where $Z_{p_1} := z^{*T}_{p_0} p^{*T}_0 P_0 Q_1 p_0 z_{p_0} \in \mathbb{R}^{n_1,n_1}$ and $Z_{q_1} := z^{*T}_{p_0} p^{*T}_0 P_0 P_1 Q_2 p_0 z_{p_0} \in \mathbb{R}^{n_1,n_1}$. We can also see that the projectors are mutually orthogonal projectors, acting in $\mathbb{R}^{n_1}$. Let $k_2 = \dim(\operatorname{im} Z_{q_1})$, and $n_2 = n_1 - k_2$, and let us consider a basis matrix $(z_{p_1}, z_{q_1}) \in \mathbb{R}^{n_1}$ made of n_2 independent columns of projection matrix Z_{p_0} and k_2 independent columns of the complementary projection matrix Z_{q_1}. We denote by $(z^*_{p_1}, z^*_{q_1})^T$ the inverse of (z_{p_1}, z_{q_1}), such that

$$z^{*T}_{p_1} z_{p_1} = I_{n_2}, \quad z^{*T}_{p_1} z_{q_1} = 0, \quad z^{*T}_{q_1} z_{p_1} = 0, \quad z^{*T}_{q_1} z_{q_1} = I_{k_2}.$$

Hence the bases of the projector products $\{Q_0, P_0Q_1, P_0P_1Q_2, P_0P_1P_2\}$ are $\{q_0, p_0 z_{q_0}, p_0 z_{p_0} z_{q_1}, p_0 z_{p_0} z_{p_1}\}$, respectively. We note that these bases form a basis matrix $(q_0, p_0 z_{q_0}, p_0 z_{p_0} z_{q_1}, p_0 z_{p_0} z_{p_1}) \in \mathbb{R}^{n,n}$ and its inverse is denoted by $(q_0, p_0 z_{q_0}, p_0 z_{p_0} z_{q_1}, p_0 z_{p_0} z_{p_1})^{-1} = (q^*_0, p^*_0 z^*_{q_0}, p^*_0 z^*_{p_0} z^*_{q_1}, p^*_0 z^*_{p_0} z^*_{p_1})^T$. Thus using (10), we can now expand x with respect to these bases, obtaining

$$x = x_P + \sum_{i=0}^{2} x_{Q,i} = p_0 z_{p_0} z_{p_1} \xi_p + q_0 \xi_{q,0} + p_0 z_{q_0} \xi_{q,1} + p_0 z_{p_0} z_{q_1} \xi_{q,2}, \tag{16}$$

where $\xi_p \in \mathbb{R}^{n_2}$, $\xi_{q,i} \in \mathbb{R}^{k_i}$, $\quad i = 0, \ldots, 2$ and with inversion expressions

$$\xi_p = z^{*T}_{p_1} z^{*T}_{p_0} p^{*T}_0 x_P, \quad \xi_{q,0} = q^{*T}_0 x_{Q,0}, \quad \xi_{q,1} = z^{*T}_{q_0} p^{*T}_0 x_{Q,1},$$
$$\xi_{q,2} = z^{*T}_{q_1} z^{*T}_{p_0} p^{*T}_0 x_{Q,2}.$$

Without loss of generality using the bases and their respective inverses the März decoupling procedure for index-3 DAEs can be reduced to a modified decoupled system

$$\xi'_p = A_p \xi_p + B_p u, \quad \xi_p(0) = z^{*T}_{p_1} z^{*T}_{p_0} p^{*T}_0 x_P(0), \tag{17a}$$
$$\xi_{q,2} = A_{q,2} \xi_p + B_{q,2} u, \tag{17b}$$
$$\xi_{q,1} = A_{q,1} \xi_p + B_{q,1} u + A_{q_{1,2}} \xi'_{q,2}, \tag{17c}$$
$$\xi_{q,0} = A_{q,0} \xi_p + B_{q,0} u + A_{q_{0,1}} \xi'_{q,1} + A_{q_{0,2}} \xi'_{q,2}, \tag{17d}$$

of dimension n, where

$$A_p := z_{p_1}^{*T} z_{p_0}^{*T} p_0^{*T} P_0 P_1 P_2 E_3^{-1} A_3 p_0 z_{p_0} z_{p_1}, \quad B_p := z_{p_1}^{*T} z_{p_0}^{*T} p_0^{*T} P_0 P_1 P_2 E_3^{-1} B,$$
$$A_{q,2} := z_{q_1}^{*T} z_{p_0}^{*T} p_0^{*T} P_0 P_1 Q_2 E_3^{-1} A_3 p_0 z_{p_0} z_{p_1}, \quad B_{q,2} := z_{q_1}^{*T} z_{p_0}^{*T} p_0^{*T} P_0 P_1 Q_2 E_3^{-1} B,$$
$$A_{q,1} := z_{q_0}^{*T} p_0^{*T} P_0 Q_1 P_2 E_3^{-1} A_3 p_0 z_{p_0} z_{p_1}, \quad B_{q,1} := z_{q_0}^{*T} p_0^{*T} P_0 Q_1 P_2 E_3^{-1} B,$$
$$A_{q,0} := q_0^{*T} Q_0 P_1 P_2 E_3^{-1} A_3 p_0 z_{p_0} z_{p_1}, \quad B_{q,0} := q_0^{*T} Q_0 P_1 P_2 E_3^{-1} B,$$
$$A_{q_{1,2}} := z_{q_0}^{*T} p_0^{*T} P_0 Q_1 Q_2 p_0 z_{p_0} z_{q_1}, \quad A_{q_{0,1}} := q_0^{*T} Q_0 Q_1 p_0 z_{q_0},$$
$$A_{q_{0,2}} := q_0^{*T} Q_0 P_1 Q_2 p_0 z_{p_0} z_{q_1}.$$

After re-arranging and simplifying this decoupled system can be simplified to a form given by:

$$\xi_p' = A_p \xi_p + B_p u, \quad \xi_p(0) = z_{p_1}^{*T} z_{p_0}^{*T} p_0^{*T} x_P(0), \tag{18a}$$

$$\xi_q = \sum_{i=0}^{2} L^i A_q A_p^i \xi_p + \sum_{i=1}^{2} \sum_{k=0}^{i-1} L^i A_q A_p^k B_p u^{(i-k-1)} + \sum_{i=0}^{2} L^i B_q u^{(i)}, \tag{18b}$$

which do not involve numerical differentiations, where $\xi_q = (\xi_{q,2}, \xi_{q,1}, \xi_{q,0})^T \in \mathbb{R}^{n-n_2}$, $A_q = (A_{q,2}, A_{q,1}, A_{q,0})^T \in \mathbb{R}^{n-n_2,n_2}$, $B_q = (B_{q,2}, B_{q,1}, B_{q,0})^T \in \mathbb{R}^{n-n_2,m}$ and $L \in \mathbb{R}^{n-n_2,n-n_2}$ is a strictly lower triangular nilpotent matrix of index 3 given by

$$L := \begin{bmatrix} 0 & 0 & 0 \\ A_{q_{1,2}} & 0 & 0 \\ A_{q_{0,2}} & A_{q_{0,1}} & 0 \end{bmatrix}. \tag{19}$$

After solving (17) or (18), we obtain the desired solution of system (1a) using the formula:

$$x = p_0 z_{p_0} z_{p_1} \xi_p + \begin{bmatrix} p_0 z_{p_0} z_{q_1} & p_0 z_{q_0} & q_0 \end{bmatrix} \xi_q. \tag{20}$$

We observed that decoupled system preserves the dimension of the DAE system as well as its stability. We can also observe that the decoupled system (17) can be written in descriptor form given by

$$\tilde{E}\xi' = \tilde{A}\xi + \tilde{B}u, \tag{21}$$

where $\tilde{E} = \begin{bmatrix} \mathrm{I} & 0 \\ 0 & -\mathrm{L} \end{bmatrix}$, $\tilde{A} = \begin{bmatrix} A_p & 0 \\ A_q & -\mathrm{I} \end{bmatrix}$, $\tilde{B} = \begin{bmatrix} B_p \\ B_q \end{bmatrix}$ and $\xi = \begin{bmatrix} \xi_p \\ \xi_q \end{bmatrix}$.
It can easily be proved that system (1a) and (21) are equivalent. It also be proved that $\det(\lambda \tilde{E} - \tilde{A}) = \det(\mathrm{I} - \lambda L)\det(\mathrm{I} - \lambda A_p)$. Since $\det(\mathrm{I} - \lambda L) = 1$. Thus,

$\sigma_f(\tilde{E}, \tilde{A}) = \sigma(A_p)$, in [2], we proved that $\sigma_f(E, A) = \sigma(A_p)$ for the case of index-2 systems but this also holds for even higher index. Hence this decoupled system inherits the mathematic properties of the DAE system.

Index-3 Systems Without a Differential Part

We now assume that $P_0 P_1 P_2 = 0$, thus (9a), reduces to

$$\mathrm{I} = P_0 P_1 Q_2 + P_0 Q_1 + Q_0, \tag{22}$$

and also the unknowns (10) reduces to

$$x_{Q,0} := Q_0 x, \quad x_{Q,1} := P_0 Q_1 x, \quad x_{Q,2} := P_0 P_1 Q_2 x. \tag{23}$$

Thus, we need to construct the bases of projectors $\{Q_0, P_0 Q_1, P_0 P_1 Q_2\}$ in order to modify the decoupling procedure presented in [6]. Repeating the same procedure as in the previous section, i.e. from (11) to (12) as follows. Then by construction, (22) can be written as, $I_{n_0} = Z_{q0} + Z_{p0}$, where $Z_{p_0} := p_0^{*T} P_0 P_1 Q_2 p_0 \in \mathbb{R}^{n_0,n_0}$, $Z_{q_0} := p_0^{*T} P_0 Q_1 p_0 \in \mathbb{R}^{n_0,n_0}$. Again, using the general form of Proposition 2.1 in [2], Z_{p_0} and Z_{q_0} are mutually orthogonal projectors, acting in $\mathbb{R}^{n_0}$. Let $k_1 = \dim(\mathrm{im}\, Z_{q_0})$, and $n_1 = n_0 - k_1$, and consider a basis matrix $(z_{p0}, z_{q0}) \in \mathbb{R}^{n_0}$ made of n_1 independent columns of projection matrix Z_{p_0} and k_1 independent columns of the complementary projection matrix Z_{q_0}. We denote by $(z_{p_0}^*, z_{q_0}^*)^T$ the inverse of (z_{p0}, z_{q0}), such that

$$z_{p0}^{*T} z_{p0} = I_{n_1}, \quad z_{p0}^{*T} z_{q0} = 0, \quad z_{q0}^{*T} z_{p0} = 0, \quad z_{q0}^{*T} z_{q0} = I_{k_1}.$$

Then the bases of the projector products $\{Q_0, P_0 Q_1, P_0 P_1 Q_2\}$ in (22) are $\{q_0, p_0 z_{q0}, p_0 z_{p0}\}$, respectively. We also note that these bases form a basis matrix $(q_0, p_0 z_{q0}, p_0 z_{p0}) \in \mathbb{R}^n$ and its inverse is denoted by $(q_0, p_0 z_{q0}, p_0 z_{p0})^{-1} = (q_0^*, p_0^* z_{q0}^*, p_0^* z_{p0}^*)^T$. Thus using (23), we can now expand x with respect to these bases, obtaining

$$x = \sum_{i=0}^{2} x_{Q,i} = p_0 z_{p0} z_{p1} \xi_p + q_0 \xi_{q,0} + p_0 z_{q0} \xi_{q,1} + p_0 z_{p0} z_{q1} \xi_{q,2}, \tag{24}$$

where $\xi_{q,i} \in \mathbb{R}^{k_i}, \quad i = 0, \ldots, 2$ and with inversion expressions

$$\xi_{q,0} = q_0^{*T} x_{Q,0}, \; \xi_{q,1} = z_{q0}^{*T} p_0^{*T} x_{Q,1}, \; \xi_{q,i} = z_{p0}^{*T} p_0^{*T} x_{Q,i}.$$

Thus using the bases and their respective inverses the März decoupling procedure for index-3 systems whose matrix pencil has no finite eigenvalues leads to modified decoupled system

$$\xi_{q,2} = B_{q,2}u, \tag{25a}$$

$$\xi_{q,1} = B_{q,1}u + A_{q_{1,2}}\xi'_{q,2}, \tag{25b}$$

$$\xi_{q,0} = B_{q,0}u + A_{q_{0,1}}\xi'_{q,1} + A_{q_{0,2}}\xi'_{q,2}, \tag{25c}$$

where

$$B_{q,2} := z_{p_0}^{*T} p_0^{*T} P_0 P_1 Q_2 E_3^{-1} B, \qquad B_{q,1} := z_{q_0}^{*T} p_0^{*T} P_0 Q_1 P_2 E_3^{-1} B,$$

$$B_{q,0} := q_0^{*T} Q_0 P_1 P_2 E_3^{-1} B,$$

$$A_{q_{1,2}} := z_{q_0}^{*T} p_0^{*T} P_0 Q_1 Q_2 p_0 z_{p_0}, \qquad A_{q_{0,1}} := q_0^{*T} Q_0 Q_1 p_0 z_{q_0},$$

$$A_{q_{0,2}} := q_0^{*T} Q_0 P_1 Q_2 p_0 z_{p_0}.$$

Also after re-arranging and simplifying this decoupled system can be written in a form given by:

$$\xi_q = \sum_{i=0}^{2} L^i B_q u^{(i)}, \tag{26}$$

where $\xi_q = (\xi_{q,2}, \xi_{q,1}, \xi_{q,0})^T \in \mathbb{R}^n$, $B_q = (B_{q,2}, B_{q,1}, B_{q,0})^T \in \mathbb{R}^{n,m}$, $L \in \mathbb{R}^{n,n}$ is a strictly lower triangular nilpotent matrix of index 3 which takes the same form as (19). After solving (25), we obtain the desired solution of system (1a) using the formula:

$$x = \begin{bmatrix} p_0 z_{p_0} & p_0 z_{q_0} & q_0 \end{bmatrix} \xi_q. \tag{27}$$

We can see that also this decoupled system also preserves the dimension of the DAE system (1a). We note that the decoupled (25) can be written in the form (21) with matrices $\tilde{E} = -\mathrm{L}$, $\tilde{A} = -\mathrm{I}$, $\tilde{B} = B_q$ and $\tilde{\xi} = \xi_q$. Hence index-3 systems can be decoupled into the form either (18) or (26). This forms reveals the hidden structure and its interconnection of the DAE system.

4 Generalized Index Aware Model Order Reduction Method

In this section, we derive the generalized index aware IMOR (GIMOR) method or IMOR-μ method which is the generalization of the IMOR method proposed in [1,2]. Although in our discussion, we focus on index-3 systems but the procedure is the same to even higher DAEs. The decoupled systems (17) and (25), we derived in the previous sections with their respective desired solution equations (20) and (27) leads to a projected control problem of (1) given by

$$\tilde{E}\xi' = \tilde{A}\xi + \tilde{B}u \tag{28a}$$

$$y = \tilde{C}^T\xi. \tag{28b}$$

where $\tilde{E}, \tilde{A} \in \mathbb{R}^{n,n}$, $\tilde{B} \in \mathbb{R}^{n,m}$ and $\tilde{C} \in \mathbb{R}^{n,\ell}$ is the projected control output matrix. $u \in \mathbb{R}^m$ and $y \in \mathbb{R}^\ell$ are as defined in system (1). We note that any system of index-μ can be transformed into the form (28) using the decoupling procedure derived in the previous section. Taking the Laplace transform of system (28) and simplifying, we obtain:

$$Y(s) = \tilde{C}^T(s\tilde{E} - \tilde{A})^{-1}\tilde{B}U(s) + \tilde{C}^T(s\tilde{E} - \tilde{A})^{-1}\tilde{E}\xi(0), \tag{29}$$

where $Y(s)$ and $U(s)$ is the Laplace transform of output $y(t)$ and input $u(t)$ respectively. If we define: $\tilde{H}(s) := \tilde{C}^T(s\tilde{E} - \tilde{A})^{-1}\tilde{B}$, as the transfer function of the projected DAE system. Then (29) can be written as:

$$Y(s) = \tilde{H}(s)U(s) + \tilde{C}^T(s\tilde{E} - \tilde{A})^{-1}\tilde{E}\xi(0). \tag{30}$$

In the model order reduction community, they assume vanishing initial condition i.e. $\xi(0) = 0$ which is not always possible for higher index DAE. This leads to the failure of existing methods to reduce higher index DAE system as illustrated in Example 1. We have already discussed that the DAE system (1) can be decoupled in two ways depending on the spectrum of the matrix pencil (E, A). Thus, we also need to derive two GIMOR methods for each case.

4.1 GIMOR Method for DAE Systems with Differential Part

In this section, we assume that the system (1a) has a matrix pencil with at least one finite eigenvalue. Thus the matrices in system (28) are defined as:

$$\tilde{E} = \begin{bmatrix} \mathrm{I} & 0 \\ 0 & -\mathrm{L} \end{bmatrix}, \quad \tilde{A} = \begin{bmatrix} A_p & 0 \\ A_q & -\mathrm{I} \end{bmatrix}, \quad \tilde{B} = \begin{bmatrix} B_p \\ B_q \end{bmatrix}, \quad \tilde{C} = \begin{bmatrix} C_p \\ C_q \end{bmatrix}, \tag{31}$$

and $\tilde{\xi} = \begin{bmatrix} \xi_p & \xi_q \end{bmatrix}^T$ is the projected state space, where L is a nilpotent matrix of index μ. Substituting the above matrices into (29) and assuming $\xi_p(0) = 0$, we obtain:

$$Y(s) := H(s)U(s) - C_q^T L\xi_q(0), \tag{32}$$

since $(\mathrm{I} - sL)^{-1}L = L$ and where the function $H(s)$ can be decomposed as: $H(s) = H_p(s) + H_q(s)$, with

$$H_p(s) = C_p^T(s\mathrm{I} - A_p)^{-1}B_p \tag{33}$$

and

$$H_q(s) = C_q^T(\mathrm{I} - sL)^{-1}\left[A_q(s\mathrm{I} - A_p)^{-1}B_p + B_q\right], \tag{34}$$

are the differential and algebraic components of this function. We note that, we can choose $\xi_p(0)$ arbitrary but not $\xi_q(0)$. For index-1 systems $L = 0 \Rightarrow Y(s) = H(s)U(s)$, thus existing methods will always work while for higher index ($\mu > 1$), $L \neq 0$, thus we cannot always prescribe the input-output relation $Y(s) = H(s)U(s)$ of the entire system. This explains why the PRIMA method failed for a index-3 system in Example 1. But for special cases when $L\xi_q(0) = 0$ even for higher index DAE system, the existing MOR methods can work and their reduced models lead to good solutions as illustrated in Example 4. In general the best way to reduce DAE system is to first split the DAE system and then reduce the differential and algebraic parts separately. This motivated us to propose the GIMOR method. Following the same steps as in [1, 2], although in a more general way we derived the GIMOR method for DAEs with differential part as follows: Using Eq. (31), we obtain the decoupled control problem of control problem (1) given by:

$$\xi_p' = A_p\xi_p + B_p u \tag{35a}$$

$$-L\xi_q' = -\xi_q + A_q\xi_p + B_q u, \tag{35b}$$

$$y = C_p^T\xi_p + C_q^T\xi_q, \tag{35c}$$

where $C_q := (C_{q,\mu-1}, \ldots, C_{q,0})^T$. If we strictly separate the above system we obtain:

$$\xi_p' = A_p\xi_p + B_p u, \tag{36a}$$

$$y_p = C_p^T\xi_p, \tag{36b}$$

and

$$-L\xi_q' = -\xi_q + A_q\xi_p + B_q u, \tag{37a}$$

$$y_q = C_q^T\xi_q, \tag{37b}$$

the differential and algebraic subsystems of the control problem (35). Thus the output equation can be reconstructed using: $y = y_p + y_q$. Thus the reduction is two steps and goes as follows:

1. Reduction of the differential part.

To reduce the differential part of the DAE system, we use control problem (36). Thus we can use any MOR techniques for ODEs. According to [11], these methods are categorized into two groups which are the Krylov subspace based methods and singular value decomposition (SVD) methods. In this paper we used the Krylov subspace based methods. As discussed in [11], We choose $s_0 \in \mathbb{C} \setminus \sigma(A_p)$ as the expansion point, and we consider order-r block Krylov subspace generated by M_p, R_p, $\mathscr{V}_p := \mathscr{K}_r(M_p, R_p) = \text{span}\{R_p, M_p R_p, \ldots, M_p^{r-1} R_p\}, \quad r \le n_{\mu-1}$, where

$$M_p := (s_0 I - A_p)^{-1}, \quad R_p := (s_0 I - A_p)^{-1} B_p. \tag{38}$$

The orthonormal basis matrix of this subspace is denoted by $V_p \in \mathbb{R}^{n_{\mu-1}, rm}$ so that $V_p^T V_p = \mathrm{I}$. We seek an approximate solution of the form $\xi_p = V_p \hat{\xi}_p$. Thus substitution ξ_p into (36) and simplifying leads to a reduced-order model of the differential part (36) given by:

$$\hat{\xi}_p' = \hat{A}_p \hat{\xi}_p + \hat{B}_p u, \tag{39a}$$

$$\hat{y}_p = \hat{C}_p^T \hat{\xi}_p, \tag{39b}$$

where $\hat{A}_p = V_p^T A_p V_p$, $\hat{B}_p = V_p^T B_p$, and $\hat{C}_p = V_p^T C_p$. The transfer function of this reduced model is given by $\hat{H}_p(s) = \hat{C}_p^T (s\mathrm{I} - \hat{A}_p)^{-1} \hat{B}_p$ which approximates (33). Thus the differential part of the DAE system is reduced but the order of the algebraic part is still unchanged.

2. Reduction of the algebraic part.

We now intend to reduce the algebraic part (37). The reduction procedure for the differential variables induces a reduction procedure for the algebraic variables. This can be shown as follows: Rewriting (37a) we obtain,

$$\xi_q = L\xi_q' + A_q \xi_p + B_q u. \tag{40}$$

Let $D = \frac{\mathrm{d}}{\mathrm{d}t}$, then (40) becomes

$$\xi_q = (\mathrm{I} - LD)^{-1} \left[A_q \xi_p + B_q u \right]. \tag{41}$$

But

$$(\mathrm{I} - LD)^{-1} = \sum_{k=0}^{\mu-1} L^k D^k, \tag{42}$$

since L is a nilpotent matrix of index μ. Substituting (42) into (41) simplifying, we obtain

$$\xi_q = \sum_{k=0}^{\mu-1} L^k (A_q \xi_p^{(k)} + B_q u^{(k)}), \tag{43}$$

where $\xi_p^{(k)} = \frac{\mathrm{d}^k \xi_p}{\mathrm{d}t^k}$ and $u^{(k)} = \frac{\mathrm{d}^k u}{\mathrm{d}t^k}$. Recalling from the previous section, the reduction of the differential part of the decoupled DAE system, which confines ξ_p to the subspace $\mathscr{V}_p$, spanned by V_p, then also $\xi_p^{(k)}$, $k = 1, \cdots, \mu - 1$, belongs to the same space. Thus from (43), we observe that for the algebraic variable ξ_q, we have the restriction

$$\xi_q \in \mathscr{V}_q = \mathscr{K}_\mu(L, b_q), \tag{44}$$

where $b_q = \begin{bmatrix} B_q & A_q V_p \end{bmatrix} \in \mathbb{R}^{n_q, r+m}$, we denote by V_q an orthonormal basis of $\mathscr{V}_q$ so that $V_q^T V_q = \mathrm{I}$. Then we can write $\xi_q = V_q \hat{\xi}_q$. We observe that the reduction from the differential part induces a reduction in the algebraic part. We know that the algebraic part has μ algebraic subsystems thus $\mathscr{V}_q$ can be partitioned as:

$$\mathscr{V}_q = \left[\mathscr{V}_{q,\mu-1}^T, \cdots, \mathscr{V}_{q,1}^T, \mathscr{V}_{q,0}^T \right]^T,$$

where the length of each partition corresponds to the block sizes of algebraic subsystems given by k_i, $i = \mu - 1, \ldots, 1, 0$, respectively. We then compute the orthonormal basis matrix of each partition to build a block diagonal orthonormal matrix:

$$\hat{V}_q = \mathrm{blkdiag}\left[V_{q,\mu-1}, \ldots, V_{q,1}, V_{q,0} \right],$$

where $V_{q,i} = \mathrm{orth}(\mathscr{V}_{q,i})$. We can now reduce the order of the algebraic part by substituting $\xi_q = V_q \hat{\xi}_q$ and $\xi_p = V_p \hat{\xi}_p$ into Eq. (37). Thus the reduced algebraic part is given by:

$$-\hat{L}\hat{\xi}_q' = -\hat{\xi}_q + \hat{A}_q \hat{\xi}_p + \hat{B}_q u, \tag{45a}$$

$$\hat{y}_q = \hat{C}_q^T \hat{\xi}_q, \tag{45b}$$

where $\hat{L} = V_q^T L V_q$, $\hat{A}_q = V_q^T A_p V_p$, $\hat{B}_q = V_q^T B_q$ and $\hat{C}_q = V_q^T C_q$.

Next, we need to show that the first moments of the transfer function $\hat{H}_q(s)$ around an arbitrary expansion point s_0 of the reduced algebraic part (45) are equal to the first moments of the transfer function (34) of the original algebraic part. This can be done as follows: Using a Neumman expansion [5], we obtain:

$$(\mathrm{I} - sL)^{-1} = \sum_{k=0}^{\mu-1} M_L^k \mathscr{P}_L (s - s_0)^k, \tag{46}$$

where

$$M_L := (\mathrm{I} - s_0 L)^{-1} L = \sum_{j=0}^{\mu-2} L^{j+1} s_0^j$$

is a nilpotent matrix of index μ at any arbitrary expansion point s_0 and

$$\mathscr{P}_L := (\mathrm{I} - s_0 L)^{-1} = \sum_{j=0}^{\mu-1} L^j s_0^j.$$

Using the same trick but with the same expansion point as in the differential part, we can expand $(s\mathrm{I} - A_p)^{-1} B_p$ as

$$(s\mathrm{I} - A_p)^{-1} B_p = \sum_{i=0}^{\infty} (-1)^i M_p^i R_p (s - s_0)^i, \tag{47}$$

where M_p and R_p as defined as in (38). Substituting Eqs. (46) and (47) into (34) and simplifying, we obtain:

$$\begin{aligned} H_q(s) = C_q^T \Bigg[& \sum_{i=0}^{\infty} \sum_{k=0}^{\mu-1} \left[(-1)^i M_L^k \mathscr{P}_L A_q M_p^i R_p (s - s_0)^{k+i} \right] \\ & + \sum_{k=0}^{\mu-1} M_L^k \mathscr{P}_L B_q (s - s_0)^k \Bigg]. \end{aligned} \tag{48}$$

Without loss of generality (48) can be written as

$$H_q(s) = \sum_{i=0}^{\infty} h_{q,i} (s - s_0)^i, \tag{49}$$

where $h_{q,k}$, $k = 0, 1, \cdots$ are the moments of the transfer function of the algebraic part $H_q(s)$. In model reduction we are always interested in the few moments of the transfer function which approximates the original transfer function. Thus, if we consider r moments of the transfer function $H_q(s)$, we obtain:

$$\hat{H}_q(s) = \sum_{i=0}^{r-1} \hat{h}_{q,i} (s - s_0)^i, \tag{50}$$

where $\hat{h}_{q,i}$, $i = 0, 1, \cdots, r-1$ are the moments of the transfer function of the reduced order algebraic part. It can be proved that $\hat{h}_{q,i}(s_0) = h_{q,i}$, $\ell = 0, 1, \ldots, r-1$.

If we combine the reduced systems (39) and (45), we obtain the reduced model of the decoupled system (35). We observe that the projected state space solution is approximated by $\xi = V\hat{\xi}$, where $V = \begin{bmatrix} V_p & 0 \\ 0 & V_q \end{bmatrix}$. Hence substituting $\xi = V\hat{\xi}$ into Eq. (28), we obtain the GIMOR model for DAE system (1) given by:

$$\hat{E}\hat{\xi}' = \hat{A}\hat{\xi} + \hat{B}u \tag{51a}$$

$$\hat{y} = \hat{C}^T\hat{\xi}, \tag{51b}$$

where $\hat{E} = V^T\tilde{E}V$, $\hat{A} = V^T\tilde{A}V$, $\hat{B} = V^T\tilde{B}$ and $\hat{C} = V^T\tilde{C}$, and its function $\hat{H}(s)$ can be written as:

$\hat{H}(s) = \hat{C}^T(s\hat{E} - \hat{A})^{-1}\hat{B}$. We note that before applying the GIMOR method, we need to first check whether some components of C_q are zeros, because if they are zero and their respective solutions are not used by other state variable. Then there respective algebraic parts can be removed.

4.2 GIMOR Method for Purely Algebraic DAE Systems

In this section, we assume that the system (1a) has a matrix pencil with only infinite eigenvalue. Thus the matrices in system (28) are defined as:

$$\tilde{E} = -\mathrm{L}, \quad \tilde{A} = -\mathrm{I}, \quad \tilde{B} = B_q, \quad \tilde{C} = C_q, \tag{52}$$

and $\tilde{\xi} = \xi_q$ is the projected state space. Substituting the above matrices into Eq. (29) and simplifying, we obtain:

$$Y(s) := \overbrace{C_q^T(\mathrm{I} - sL)^{-1}B_q}^{H(s)}\,U(s) - C_q^T L\xi_q(0).$$

Thus the $H(s)$ function of the DAE system (1) is equal to that of the algebraic part $H_q(s)$, i.e. $H(s) = H_q(s)$. Thus the GIMOR method can be derived directly from step 2 of the previous section ignoring the differential components. From Eq. (44), we derive the subspace:

$$\mathscr{V}_q := \mathscr{K}_\mu(L, B_q) = \operatorname{span}\{B_q, LB_q, \ldots, L^{\mu-1}B_q\}, \tag{53}$$

which we can use to reduce system (52). This subspace can also be partitioned into μ algebraic parts as in the previous case. We can then compute the orthonormal basis matrix of each partition to build a block diagonal orthonormal matrix:

$$V_q = \operatorname{blkdiag}\left[V_{q,\mu-1}, \ldots, V_{q,1}, V_{q,0}\right],$$

where $V_{q,i} = \mathrm{orth}(\mathscr{V}_{q,i})$. We then reduce the order of the DAE system by substituting $\xi_q = V_q\hat{\xi}_q$ into Eq. (52). Thus the reduced decoupled system is given by:

$$-\hat{L}\hat{\xi}_q' = -\hat{\xi}_q + \hat{B}_q u, \tag{54a}$$

$$\hat{y}_q = \hat{C}_q^T \hat{\xi}_q, \tag{54b}$$

where $\hat{L} = V_q^T L V_q$, $\hat{B}_q = V_q^T B_q$ and $\hat{C}_q = V_q^T C_q$. We also observe that the projected state space solution is approximated by $\xi = V\hat{\xi}$, where $V = \hat{V}_q$. Hence substituting $\xi = V\hat{\xi}$ into Eq. (28), we obtain the GIMOR model for DAE system (1) given by:

$$\hat{E}\hat{\xi}' = \hat{A}\hat{\xi} + \hat{B}u \tag{55a}$$

$$\hat{y} = \hat{C}^T\hat{\xi}, \tag{55b}$$

where $\hat{E} = V^T\tilde{E}V$, $\hat{A} = V^T\tilde{A}V$, $\hat{B} = V^T\tilde{B}$ and $\hat{C} = V^T\tilde{C}$. The transfer function of the GIMOR model is given by $\hat{H}(s) = \hat{H}_q(s) = \hat{C}^T(s\hat{E} - \hat{A})^{-1}\hat{B} = \hat{C}_q^T(\mathrm{I} - s\hat{L})^{-1}\hat{B}_q$.

Remark 1 We note that the GIMOR method also satisfies the goals of model order reduction such as; (i) the approximation error must be small, (ii) properties of the original system such as stability and passivity are preserved, and (iii) the reduction procedure is computationally efficient If we focus on the DAEs with matrix pair (E, A) whose finite spectrum has at least one finite eigenvalue. We have discussed that the $\sigma_f(E, A) = \sigma(A_p)$, thus the stability and passivity of the reduced-order model of the GIMOR method depends on the MOR method used to reduce the differential part. In this paper, we use the PRIMA method which always guarantees these two properties for the case of SISO systems. It can also easily be proved that if the first r moments of $H_p(s)$ around $s = s_0$ are preserved, then we can ensure that the first r moments of $H_q(s)$ around the same expansion point are also preserved by the GIMOR methods. Thus the GIMOR method preserve the moments of function $H(s)$. In the next section, we illustrate the GIMOR method leads to simple sparse models which are accurate and computationally cheap. In additional it always preserve the index of the DAE system. Thus the GIMOR method satisfies the goals of the model order reduction.

5 Numerical Experiments

In this section, we present some simple and large examples that illustrate the robustness of the GIMOR reduced-order models as compared to PRIMA reduced-order models. We note that we use the PRIMA method to reduce the differential part in the GIMOR methods but also other methods can be used.

5.1 Simple Examples

Example 2 In this example, we use system matrices $(E, \ A, \ B, \ C)$ from Example 1. The matrix pencil $(E, \ A)$ of this system has only one finite eigenvalue, thus its decoupled system must have only one differential equation and eight algebraic equations. In order to apply the GIMOR method, we need to first decouple the system into differential and algebraic parts. The system can be decoupled into the form (35). First, we need to determine the tractability index of system (3) using the formula (5) and then construct special projectors such that (7) holds. The matrix pencil $(E, \ A)$ has projectors and matrix chains given by:

$$Q_0 = \begin{bmatrix} 1 & 0 & 1 & 0 & -1 & 0 & 0 & 0 & 0 \\ 0 & 1 & 0 & 1 & 1 & 0 & 0 & 0 & 0 \\ 0 & 0 & 0 & 0 & 0 & 0 & 0 & 0 & 0 \\ 0 & 0 & 0 & 0 & 0 & 0 & 0 & 0 & 0 \\ 0 & 0 & 0 & 0 & 0 & 0 & 0 & 0 & 0 \\ 0 & 0 & 0 & -1 & 0 & 1 & 0 & 0 & 0 \\ 0 & 0 & 0 & 1 & 2 & 0 & 1 & 0 & 0 \\ 0 & 0 & 0 & -1 & -1 & 0 & 0 & 1 & 0 \\ 0 & 0 & 0 & -1 & -1 & 0 & 0 & 0 & 1 \end{bmatrix}, \quad Q_1 = \begin{bmatrix} 0 & 0 & 0 & 0 & 0 & 0 & 0 & 0 & 0 \\ 0 & 0 & 0 & 0 & 0 & 0 & 0 & 0 & 0 \\ 0 & 0 & 1 & 1 & 0 & 0 & 0 & 0 & 0 \\ 0 & 0 & 0 & 0 & 0 & 0 & 0 & 0 & 0 \\ 0 & 0 & 0 & 0 & 0 & 0 & 0 & 0 & 0 \\ 0 & 0 & 0 & 0 & 0 & 0 & 0 & 0 & 0 \\ 0 & 0 & 0 & 0 & 0 & 0 & 0 & 0 & 0 \\ 0 & 0 & 0 & 0 & 0 & 0 & 0 & 0 & 0 \\ 0 & 0 & 0 & 0 & 0 & 0 & 0 & 0 & 0 \end{bmatrix},$$

$$Q_2 = \begin{bmatrix} 0 & 0 & 0 & 0 & 0 & 0 & 0 & 0 & 0 \\ 0 & 0 & 0 & 0 & 0 & 0 & 0 & 0 & 0 \\ 0 & 0 & 0 & 0 & 0 & 0 & 0 & 0 & 0 \\ 0 & 0 & 0 & 1 & 0 & 0 & 0 & 0 & 0 \\ 0 & 0 & 0 & -1 & 0 & 0 & 0 & 0 & 0 \\ 0 & 0 & 0 & 0 & 0 & 0 & 0 & 0 & 0 \\ 0 & 0 & 0 & 0 & 0 & 0 & 0 & 0 & 0 \\ 0 & 0 & 0 & 0 & 0 & 0 & 0 & 0 & 0 \\ 0 & 0 & 0 & 0 & 0 & 0 & 0 & 0 & 0 \end{bmatrix}$$

$$\text{and} \quad P_i = \mathrm{I} - Q_i, \ i = 0, 1, 2, \tag{56}$$

and the last matrix chains are given by:

$$E_3 = \begin{bmatrix} 0 & 0 & 0 & 1 & 0 & 0 & 0 & 0 & 0 \\ 0 & 0 & -1 & 1 & 0 & 0 & 0 & 0 & 0 \\ -1 & -1 & 1 & 0 & 0 & 0 & 0 & 0 & 0 \\ 0 & 1 & 0 & 0 & 1 & 0 & 0 & 0 & 0 \\ 0 & 0 & 1 & 0 & 0 & 1 & 0 & 0 & 0 \\ 0 & 0 & 0 & -2 & -1 & 0 & -1 & 0 & 0 \\ 0 & 0 & 0 & 0 & -1 & 0 & 0 & 1 & 0 \\ 0 & 0 & 0 & 0 & -1 & 0 & 0 & 0 & 1 \\ 0 & 0 & 0 & 0 & 0 & 1 & -1 & -1 & -1 \end{bmatrix}, \ A_3 = \begin{bmatrix} 0 & 0 & 0 & 0 & 0 & 0 & 0 & 0 & 0 \\ 0 & 0 & 0 & 0 & 0 & 0 & 0 & 0 & 0 \\ 0 & 0 & 0 & 0 & 0 & 0 & 0 & 0 & 0 \\ 0 & 0 & 0 & 0 & 0 & 0 & 0 & 0 & 0 \\ 0 & 0 & 0 & 0 & 0 & 0 & 0 & 0 & 0 \\ 0 & 0 & 0 & -2 & -2 & 0 & 0 & 0 & 0 \\ 0 & 0 & 0 & 0 & 0 & 0 & 0 & 0 & 0 \\ 0 & 0 & 0 & 0 & 0 & 0 & 0 & 0 & 0 \\ 0 & 0 & 0 & 0 & 0 & 0 & 0 & 0 & 0 \end{bmatrix}, \tag{57}$$

respectively. Since E_3 is non-singular thus this is an index-3 system. Thus its decoupled system must take the form (17) or (18). Next, we use the matrix and projectors chains (56) and (57) to decouple the DAE system using the method presented in Sect. 3.1 as follows: We construct basis column matrices and their respective inverse for projector Q_0 and P_0 given by:

$$q_0 = \begin{bmatrix} 1 & 0 & 0 & 0 & 0 & 0 \\ 0 & 1 & 0 & 0 & 0 & 0 \\ 0 & 0 & 0 & 0 & 0 & 0 \\ 0 & 0 & 0 & 0 & 0 & 0 \\ 0 & 0 & 0 & 0 & 0 & 0 \\ 0 & 0 & 1 & 0 & 0 & 0 \\ 0 & 0 & 0 & 1 & 0 & 0 \\ 0 & 0 & 0 & 0 & 1 & 0 \\ 0 & 0 & 0 & 0 & 0 & 1 \end{bmatrix}, \ p_0 = \begin{bmatrix} -1 & 0 & 1 \\ 0 & -1 & -1 \\ 1 & 0 & 0 \\ 0 & 1 & 0 \\ 0 & 0 & 1 \\ 0 & 1 & 0 \\ 0 & -1 & -2 \\ 0 & 1 & 1 \\ 0 & 1 & 1 \end{bmatrix} \ \text{and} \ q_0^{*T} = \begin{bmatrix} 1 & 0 & 0 & 0 & 0 & 0 \\ 0 & 1 & 0 & 0 & 0 & 0 \\ 1 & 0 & 0 & 0 & 0 & 0 \\ 0 & 1 & -1 & 1 & -1 & -1 \\ -1 & 1 & 0 & 2 & -1 & -1 \\ 0 & 0 & 1 & 0 & 0 & 0 \\ 0 & 0 & 0 & 1 & 0 & 0 \\ 0 & 0 & 0 & 0 & 1 & 0 \\ 0 & 0 & 0 & 0 & 0 & 1 \end{bmatrix}^T,$$

$$p_0^{*T} = \begin{bmatrix} 0 & 0 & 1 & 0 & 0 & 0 & 0 & 0 & 0 \\ 0 & 0 & 0 & 1 & 0 & 0 & 0 & 0 & 0 \\ 0 & 0 & 0 & 0 & 1 & 0 & 0 & 0 & 0 \end{bmatrix}.$$

Next, we use the above basis matrices to construct projectors

$$Z_{p_0} = p_0^{*T}\,[P_0 P_1 P_2 + P_0 P_1 Q_2]\,p_0 = \begin{bmatrix} 0 & -1 & 0 \\ 0 & 1 & 0 \\ 0 & 0 & 1 \end{bmatrix} \quad \text{and}$$

$$Z_{q_0} = p_0^{*T} P_0 Q_1 p_0 = \begin{bmatrix} 1 & 1 & 0 \\ 0 & 0 & 0 \\ 0 & 0 & 0 \end{bmatrix},$$

which are orthogonal complement to each other. We construct their basis column matrices and respective inverses given by:

$$z_{p0} = \begin{bmatrix} -1 & 0 \\ 1 & 0 \\ 0 & 1 \end{bmatrix}, \; z_{q0} = \begin{bmatrix} 1 \\ 0 \\ 0 \end{bmatrix} \quad \text{and} \quad z_{p0}^{*T} = \begin{bmatrix} 0 & 1 & 0 \\ 0 & 0 & 1 \end{bmatrix}, \quad z_{q0}^{*T} = \begin{bmatrix} 1 & 1 & 0 \end{bmatrix}.$$

Then, use the above matrices to construct another projectors given by:

$$Z_{p1} = z_{p0}^{*T} p_0^{*T} P_0 P_1 P_2 p_0 z_{p0} = \begin{bmatrix} 0 & 0 \\ 1 & 1 \end{bmatrix} \quad \text{and}$$

$$Z_{q1} = z_{p0}^{*T} p_0^{*T} P_0 P_1 Q_2 p_0 z_{p0} = \begin{bmatrix} 1 & 0 \\ -1 & 0 \end{bmatrix}.$$

and their respective basis column and inverse matrices given by:

$$z_{p1} = \begin{bmatrix} 0 \\ 1 \end{bmatrix}, \; z_{q1} = \begin{bmatrix} 1 \\ -1 \end{bmatrix} \quad \text{and} \quad z_{p1}^{*T} = \begin{bmatrix} 1 & 0 \end{bmatrix}, \quad z_{q1}^{*T} = \begin{bmatrix} 1 & 1 \end{bmatrix}. \tag{58}$$

Recall we have already derived that decoupled system for index-3 system in Sect. 3.1. Thus substituting Eqs. (56)–(58) into (17), we obtain a decoupled system of the form (17) with system matrices:

$$A_p = -2, \quad B_p = 2, \quad A_{q,2} = 0, \quad B_{q,2} = 1, \quad A_{q,1} = 0, \quad B_{q,1} = 1, \quad A_{q1,2} = 1,$$

$$A_{q,0} = \begin{bmatrix} 0 \\ 0 \\ 0 \\ 0 \\ 0 \\ 0 \end{bmatrix}, \quad B_{q,0} = \begin{bmatrix} 0 \\ 1 \\ -1 \\ 1 \\ -1 \\ -1 \end{bmatrix}, \quad A_{q0,1} = \begin{bmatrix} 1 \\ 0 \\ 0 \\ 0 \\ 0 \\ 0 \end{bmatrix} \quad \text{and} \quad A_{q0,2} = \begin{bmatrix} 0 \\ 0 \\ -1 \\ -1 \\ 0 \\ 0 \end{bmatrix}. \tag{59}$$

Hence the DAE system is decoupled into $n_p = 1$ and $k_2 + k_1 + k_0 = 1 + 1 + 6 = 8$ differential and algebraic parts as expected. This decoupled system can further be simplified into a decoupled system in the form (18) with system matrices:

$$L = \begin{bmatrix} 0&0&0&0&0&0&0&0&0 \\ 1&0&0&0&0&0&0&0&0 \\ 0&1&0&0&0&0&0&0&0 \\ 0&0&0&0&0&0&0&0&0 \\ -1&0&0&0&0&0&0&0&0 \\ -1&0&0&0&0&0&0&0&0 \\ 0&0&0&0&0&0&0&0&0 \\ 0&0&0&0&0&0&0&0&0 \end{bmatrix}, \quad A_q = \begin{bmatrix} 0 \\ 0 \\ 0 \\ 0 \\ 0 \\ 0 \\ 0 \\ 0 \end{bmatrix}, \quad B_q = \begin{bmatrix} 1 \\ 1 \\ 0 \\ 1 \\ -1 \\ 1 \\ -1 \\ -1 \end{bmatrix}.$$

Hence the decoupled system of (3) is given by:

$$\xi_p' = -2\xi_p + 2u, \quad \xi_p(0) = \phi(0) + \mathrm{I}(0) \tag{60a}$$

$$\xi_q = \sum_{i=0}^{2} L^i B_q u^{(i)}. \tag{60b}$$

System (60) can be solved hierarchically and using Eq. (20), we can compute the desired output solution $y = C^T x$ of the DAE system (3) given by:

$$y = \xi_p + \begin{bmatrix} 0 & 0 & 0 & 0 & 0 & 0 & 0 & 1 \end{bmatrix} \xi_q.$$

Hence the desired solution of the DAE system (3) is given by:

$$y = C^T x = \xi_p(0)e^{-2t} - 2e^{-2t} \int_0^t e^{2\tau} u \, \mathrm{d}\tau - u,$$

which coincides with the exact solution of the DAE system. We observe that decouple system (60) is much easier to solve than solving DAE system (3). It can easily be checked that $L\xi_q(0) \neq 0$, this explains the reason why the PRIMA method failed to reduce this example as illustrated in Example 1. Next, we need to reduce the order of the DAE system (3) while preserving its original properties. First we derive the reduced-order model of system (3) using the GIMOR method. We need to first write the decoupled system (60) into descriptor form given by:

$$\tilde{E} = \begin{bmatrix} 1 & 0 & 0 & 0 & 0 & 0 & 0 & 0 & 0 \\ 0 & 0 & 0 & 0 & 0 & 0 & 0 & 0 & 0 \\ 0 & -1 & 0 & 0 & 0 & 0 & 0 & 0 & 0 \\ 0 & 0 & -1 & 0 & 0 & 0 & 0 & 0 & 0 \\ 0 & 0 & 0 & 0 & 0 & 0 & 0 & 0 & 0 \\ 0 & 1 & 0 & 0 & 0 & 0 & 0 & 0 & 0 \\ 0 & 1 & 0 & 0 & 0 & 0 & 0 & 0 & 0 \\ 0 & 0 & 0 & 0 & 0 & 0 & 0 & 0 & 0 \\ 0 & 0 & 0 & 0 & 0 & 0 & 0 & 0 & 0 \end{bmatrix}, \quad \tilde{A} = \begin{bmatrix} -2 & 0 & 0 & 0 & 0 & 0 & 0 & 0 & 0 \\ 0 & -1 & 0 & 0 & 0 & 0 & 0 & 0 & 0 \\ 0 & 0 & -1 & 0 & 0 & 0 & 0 & 0 & 0 \\ 0 & 0 & 0 & -1 & 0 & 0 & 0 & 0 & 0 \\ 0 & 0 & 0 & 0 & -1 & 0 & 0 & 0 & 0 \\ 0 & 0 & 0 & 0 & 0 & -1 & 0 & 0 & 0 \\ 0 & 0 & 0 & 0 & 0 & 0 & -1 & 0 & 0 \\ 0 & 0 & 0 & 0 & 0 & 0 & 0 & -1 & 0 \\ 0 & 0 & 0 & 0 & 0 & 0 & 0 & 0 & -1 \end{bmatrix}$$

$$\tilde{B} = \begin{bmatrix} 2 & 1 & 1 & 0 & 1 & -1 & 1 & -1 & -1 \end{bmatrix}^T, \quad \tilde{C} = \begin{bmatrix} 1 & 0 & 0 & 0 & 0 & 0 & 0 & 0 & 1 \end{bmatrix}^T. \tag{61}$$

The transfer functions of the DAE systems (3) and (61) must coincide, which is given by:

$$H(s) = \tilde{H}(s) = -\frac{s}{s+2}. \tag{62}$$

Using $s_0 = 0$ as the expansion point and following the procedure in Sect. 4.1 we can construct the block orthonormal basis matrix given by:

$$V = \begin{bmatrix} 1\,0\,0\,0 & 0 & 0 & 0 & 0 & 0 \\ 0\,1\,0\,0 & 0 & 0 & 0 & 0 & 0 \\ 0\,0\,1\,0 & 0 & 0 & 0 & 0 & 0 \\ 0\,0\,0\,0 & -0.4472 & 0.4472 & -0.4472 & 0.4472 & 0.4472 \end{bmatrix}^T,$$

of the system (61). Thus the GIMOR reduced-order model of DAE system (3) is given by:

$$\tilde{E}_r = V^T \tilde{E} V = \begin{bmatrix} 1 & 0\,0\,0 \\ 0 & 0\,0\,0 \\ 0 & -1\,0\,0 \\ 0 & 0\,0\,0 \end{bmatrix}, \quad \tilde{A}_r = V^T \tilde{A} V = \begin{bmatrix} -2 & 0 & 0 & 0 \\ 0 & -1 & 0 & 0 \\ 0 & 0 & -1 & 0 \\ 0 & 0 & 0 & -1 \end{bmatrix}$$

$$\tilde{B}_r = V^T \tilde{B} = \begin{bmatrix} 2 \\ 1 \\ 1 \\ -2.2361 \end{bmatrix}, \quad \text{and} \quad \tilde{C}_r = V^T \tilde{C} = \begin{bmatrix} 1 \\ 0 \\ 0 \\ 0.4472 \end{bmatrix}^T. \tag{63}$$

Thus system (3) is reduced from order 9 to order 4 using GIMOR model. The next step is to check the validity of the reduced model. We observed that the transfer function and exact solution of the GIMOR model is given by:

$$\tilde{H}_r(s) = \frac{2}{s+2} - 1 \quad \text{and} \quad \tilde{y}_r(s) = \xi_p(0)e^{-2t} - 2e^{-2t} \int_0^t e^{2\tau} u \, \mathrm{d}\tau - u,$$

respectively, which coincides with that of the original model (3). The reduced model also preserves the stability of the original system, i.e. $\sigma_f(E, A) = \sigma_f(\tilde{E}_r, \tilde{A}_r) = \{-2\}$. Hence, the reduced system (63) is a good reduced-order model for DAE system (3). We conclude that the GIMOR method leads to a good reduced-order model while the PRIMA method failed to solve this system.

Example 3 This example originates from [10]. We consider a Miller integrator electric network. Using modified nodal analysis on this electric network leads to a DAE system with system matrices:

$$E = \begin{bmatrix} 0 & 0 & 0 & 0\,0 \\ 0 & C_1 + C_2 & -C_2 & 0\,0 \\ 0 & -C_2 & C_2 & 0\,0 \\ 0 & 0 & 0 & 0\,0 \\ 0 & 0 & 0 & 0\,0 \end{bmatrix}, B = \begin{bmatrix} 0 \\ 0 \\ 0 \\ 1 \\ 0 \end{bmatrix}, C = \begin{bmatrix} 0 \\ 0 \\ 1 \\ 0 \\ 0 \end{bmatrix}, A = \begin{bmatrix} -G_1 & G_1 & 0 & -1 & 0 \\ G_1 & -G_1 & 0 & 0 & 0 \\ 0 & 0 & 0 & 0 & -1 \\ -1 & 0 & 0 & 0 & 0 \\ 0 & a & -1 & 0 & 0 \end{bmatrix},$$

where $G_1 = \frac{1}{R_1} > 0$ and the state vector is given by $x = [e_1, e_2, e_3, j_{v_1}, j_{v_2}]^T$. This system is solvable since its matrix pencil is regular, i.e.,

$$\det(\lambda E - A) = [C_1 + C_2(1-a)]\,\lambda + G_1 \neq 0.$$

According to [10], the system is of index 3 if $a = 1+\frac{C_1}{C_2}$. This implies that $\det(\lambda E - A) = G_1 \in \mathbb{C} \setminus \{0\}$ and the resulting systems leads to a purely algebraic system. Thus the decoupled system of the DAE system (64) takes the form (25). Let $C_1 = 10^{-8}$, $C_2 = 10^{-8}$, $R_1 = 10^3$ and $a = 2$, then its decoupled system has the matrices

$$B_{q,2} = 5.389 \times 10^{-13}, \qquad B_{q,1} = 2,$$

$$A_{q1,2} = -4.9999 \times 10^{-5}, \qquad B_{q,0} = \begin{bmatrix} 1 \\ 1.2434 \times 10^{-16} \\ -10^{-8} \end{bmatrix}$$

$$A_{q0,1} = \begin{bmatrix} -1.7151 \times 10^{-19} \\ -6.2057 \times 10^{-17} \\ -5 \times 10^{-9} \end{bmatrix}, \qquad A_{q0,2} = \begin{bmatrix} -1.6153 \times 10^{-19} \\ -2.4999 \times 10^{-8} \\ 1.5 \times 10^{-8} \end{bmatrix}$$

and

$$C_q = \left[-0.49994\ 1\ 0\ 0\ 0\right]^T.$$

Then the decoupled system can be written into the form (26) with system matrices:

$$L := \begin{bmatrix} 0 & 0 & 0\,0\,0 \\ -4.9999 \times 10^{-5} & 0 & 0\,0\,0 \\ -1.6153 \times 10^{-19} & -1.7151 \times 10^{-19} & 0\,0\,0 \\ -2.4999 \times 10^{-8} & -6.2057 \times 10^{-17} & 0\,0\,0 \\ 1.5 \times 10^{-8} & -5 \times 10^{-9} & 0\,0\,0 \end{bmatrix}, \quad B_q := \begin{bmatrix} 5.389 \times 10^{-13} \\ 2 \\ 1 \\ 1.2434 \times 10^{-16} \\ -1 \times 10^{-8} \end{bmatrix}.$$

It can easily be seen that L is a lower triangular nilpotent matrix of index 3. Hence the solution of the DAE system (64) is given by:

$$y = C_q^T \sum_{i=0}^{2} L^i B_q u^{(i)} = 2u - 2.6944 \times 10^{-17} u' \approx 2u.$$

We observe that solving the decoupled system is far easier to solve than solving the original DAE system. We also observed that $L\xi_q(0) \neq 0$, thus we expect the PRIMA method to fail as explained in Sect. 4. The next step is to reduce the order of the DAE system (64) using GIMOR and PRIMA methods. We first used the GIMOR method: The decoupled system in the descriptor form can be written as:

$$\tilde{E} = -\mathrm{L}, \quad \tilde{A} = -\mathrm{I}, \quad \tilde{B} = B_q, \quad \tilde{C} = C_q, \tag{64}$$

Since the decoupled system is purely algebraic, we need to follow procedure in Sect. (4.2). Thus we able to construct a diagonal orthonormal basis matrix given by: $V = \begin{bmatrix} 1\,0\,0\,0\,0 \\ 0\,1\,0\,0\,0 \end{bmatrix}^T$. Thus the GIMOR reduced-order model is given by:

$$\tilde{E}_r = V^T \tilde{E} V = \begin{bmatrix} 0 & 0 \\ -4.9999 \times 10^{-5} & 0 \end{bmatrix}, \quad \tilde{A}_r = V^T \tilde{A} V = \begin{bmatrix} -1 & 0 \\ 0 & -1 \end{bmatrix},$$

$$\tilde{B}_r = V^T \tilde{B} = \begin{bmatrix} 5.389 \times 10^{-13} \\ 2 \end{bmatrix} \quad \text{and} \quad \tilde{C}_r = V^T \tilde{C} = \begin{bmatrix} -0.49994 \\ 1 \end{bmatrix}. \tag{65}$$

We observed that the solution of the GIMOR model coincides with that of the original model (64). Thus the reduced-order model (65) is a good reduced model for DAE system (64). For comparison, we also constructed another reduced-order model of (64) using the using PRIMA method [8]. Using the $s_0 = 0$ as the expansion point, we were able construct its orthonormal basis matrix V_r given by: $V_r = \begin{bmatrix} 0.40825 & 0.40825 & 0.8165 & 0 & 0 \\ 0 & 0 & 0 & 0 & 1 \end{bmatrix}^T$. We used V_r to construct PRIMA reduced-order model of (64) given by:

$$E_r = V_r^T E V_r = \begin{bmatrix} 3.3333 \times 10^{-9} & 0 \\ 0 & 0 \end{bmatrix},$$

$$A_r = V_r^T A V_r = \begin{bmatrix} [r] - 2.4074 \times 10^{-35} & -0.8165 \\ 0 & 0 \end{bmatrix}, \tag{66}$$

$$B_r = V_r^T B = \begin{bmatrix} 0 \\ 0 \end{bmatrix} \quad \text{and} \quad C_r = V_r^T C = \begin{bmatrix} 0.8165 \\ 0 \end{bmatrix}.$$

We observed that the PRIMA model is unsolvable since $\det(\lambda E_r - A_r) = 0$. Clearly, the PRIMA method still fails for this example while the GIMOR method leads to good reduced-order model.

5.2 Large Examples

Example 4 In this example we consider a Constrained Damped Mass-Spring System as described in [7]. The vibration of this system is described by a DAE system with system matrices:

$$E = \begin{bmatrix} I & 0 & 0 \\ 0 & M & 0 \\ 0 & 0 & 0 \end{bmatrix}, \quad A = \begin{bmatrix} 0 & I & 0 \\ K & D & -G^T \\ G & 0 & 0 \end{bmatrix}, \quad B = \begin{bmatrix} 0 \\ B_2 \\ 0 \end{bmatrix}, \quad C = \begin{bmatrix} C_1 \\ 0 \\ 0 \end{bmatrix}, \tag{67}$$

where $M = \text{diag}(m_1, \cdots, m_g)$ is the mass matrix, D and K are tridiagonal damping and stiffness matrices, $G = [1, 0, \cdots, 0, -1] \in \mathbb{R}^{1,g}$ is the constraint matrix, $B_2 = e_1$ and $C_1 = \left[e_1, e_2, e_{g-1}\right]$. Here e_i denotes the ith column of the identity matrix I_g. The state variables are given by $\mathbf{x}(t) = [\mathbf{p}(t), \mathbf{v}(t), \boldsymbol{\lambda}(t)]^T$, where $\mathbf{p}(t) \in \mathbb{R}^g$ is the position vector, $\mathbf{v}(t) \in \mathbb{R}^g$ is the velocity vector and, $\boldsymbol{\lambda}(t) \in \mathbb{R}$ is the Lagrange multiplier. Thus the system is of dimension $n = 2g + 1$. The DAE system (67) is of index 3. We used the same constants as used in [7] to generate the system matrices. For $g = 6{,}000$, we obtain a system of order $n = 12{,}001$ with 1 input and 3 outputs.

This system is solvable and its matrix pencil has 11,998 finite and 3 infinite eigenvalues. Thus, its decoupled system must have 11,998 differential equations and 3 algebraic equations. For this example $L\xi_q(0) = 0$, thus the conventional MOR methods can also be used. We used $s_0 = 10^{-4}$ as the expansion point and applied our proposed GIMOR method this system and obtained a reduced-order model as shown in the Table 1. We can see that the original model is reduced to a GIMOR model of total dimension 9. In fact 99.99 % of the original model is reduced.

For comparison we also reduced the DAE system using the PRIMA method to the same dimension. In Fig. 2 we validate the reduced-order models of the two methods by comparing their respective magnitude of transfer functions with that the transfer function of the original model. In Fig. 2a we observe that the magnitude of the transfer function of the reduced-order models coincides with the original model with very small error at lower frequencies as shown in Fig. 2b.

In Fig. 3 we observe that the solutions of the reduced-order model coincides with the original model. We note that the solutions $y_2(t)$ and $y_3(t)$ coincide.

Table 1 Dimension of Original and GIMOR model

Models	Dimension # Differential eqns	# Algebraic eqns
Original model	11,998	3
GIMOR model	8	1

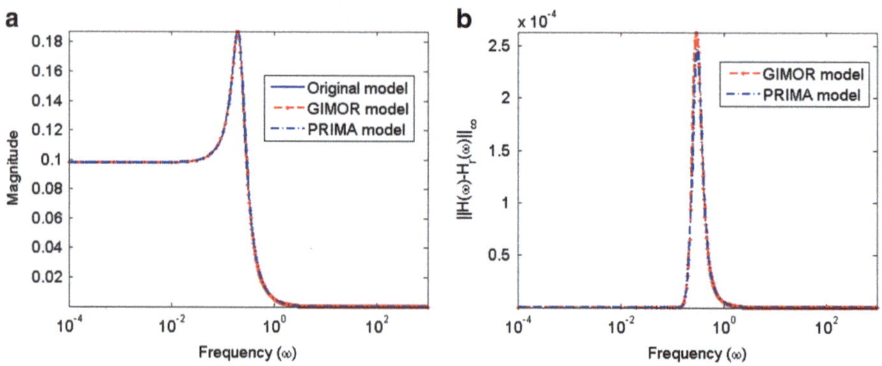

Fig. 2 Comparison of the frequency response and its error. (**a**) Frequency response. (**b**) Frequency response error

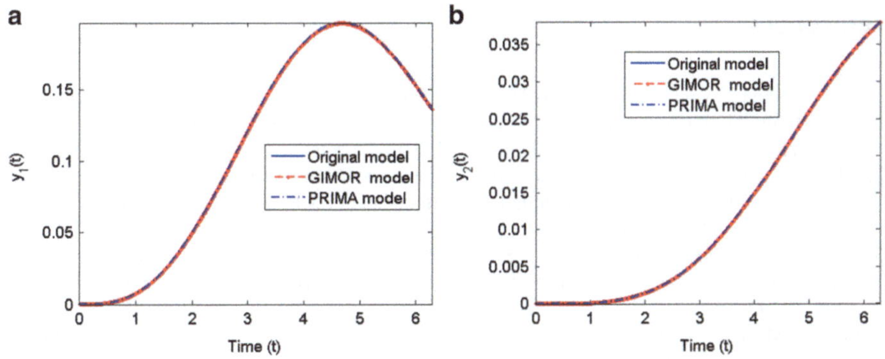

Fig. 3 Solutions using $u(t) = 10\sin(t)$ as the input function. (**a**) Solution $y_1(t)$. (**b**) Solutions $y_2(t)$

Conclusion

We have discussed that DAEs can be decoupled into differential and algebraic parts using special basis vectors of projectors and the resulting decoupled system preserve the mathematical properties of the system. Then the derived decoupled system can be used to derive the GIMOR method which is more reliable than the conventional MOR methods for reducing DAEs. For example, we have illustrated using an example that the PRIMA method leads to unsolvable reduced-order models when applied to some index-3 systems. Hence the conventional MOR methods cannot be used to reduce higher index DAEs in general. We have proposed a method for reducing higher index DAEs which is more reliable and accurate than the existing methods which we call the GIMOR method.

The DAEs are not just difficult to reduced but also to solve thus the decoupling technique can be used to solve DAEs accurately. However, one has to be aware that the numerical computation of the bases for the decoupling may involve serious difficulties because of the accuracy sensitive rank decisions [2]. But this can be avoided by computing the bases in a robust way using the structure of the matrix pair (E, A).

Acknowledgements We would like to thank our collaborators G. Alì and C. Tischendorf for their contribution to this work. This work was supported by The Netherlands Organisation for Scientific Research (NWO).

References

1. Alì, G., Banagaaya, N., Schilders, W.H.A., Tischendorf, C.: Index-aware model order reduction for differential-algebraic equations. Math. Comput. Model. Dyn. Syst.: Methods Tools Appl. Eng. Relat. Sci. (2013). doi:10.1080/13873954.2013.829501
2. Alì, G., Banagaaya, N., Schilders, W.H.A., Tischendorf, C.: Index-aware model order reduction for index-2 differential-algebraic equations with constant coefficients. SIAM J. SCI. Comput. **35**(3), A1487–A1510 (2013)
3. Freitas, F.D., Martins, N., Varrichio, S.L., Rommes, J., Véliz, F.C.V.: Reduced-order tranfer matrices from network descriptor models of electric power grids. IEEE Trans. Power Syst. **26**, 1905–1919 (2005)
4. Gerdin, M.: Parameter estimation in linear descriptor systems, PhD thesis, Linköping University, Linköping (2004)
5. Grimme, E.J.: Krylov projection methods for model reduction, PhD thesis, University of Illinois at Urbana-Champaign (1997)
6. März, R.: Canonical projectors for linear differential algebraic equations. Comput. Math. Appl. **31**(4/5), 121–135 (1996)
7. Mehrmann, V., Stykel, T.: Balanced truncation model reduction for large scale systems in descriptor form. In: Dimension Reduction of Large Scale Systems. Lecture Notes in Computational Science and Engineering, vol. 45, pp. 83–115. Springer, Berlin/New York (2005)
8. Odabasioglu, A., Celik, M., Pileggi, L.T.: PRIMA: passive reduced-order interconnect macromodeling algorithm. IEEE Trans. Comput.-Aided Des. **17**(8), 645–654 (1998)
9. Reis, T., Stykel, T.: Positive real and bounded real balancing for model reduction of descriptor systems. Int. J. Control **83**, 74–88 (2010)
10. Schilders, W.H.A., ter Maten, E., Jan, W. (eds.): Handbook of Numerical Analysis, vol. 13. Elsevier, Amsterdam/Boston (2005)
11. Schilders, W.H.A., Van der Vorst, H., Rommes, J.: Model Order Reduction: Theory, Research Aspects and Applications. Springer, Berlin/Heidelberg (2008)

Model Order Reduction of Differential Algebraic Equations Arising from the Simulation of Gas Transport Networks

Sara Grundel, Lennart Jansen, Nils Hornung, Tanja Clees, Caren Tischendorf, and Peter Benner

Abstract We explore the Tractability Index of Differential Algebraic Equations (DAEs) that emerge in the simulation of gas transport networks. Depending on the complexity of the network, systems of index 1 or index 2 can arise. It is then shown that these systems can be rewritten as Ordinary Differential Equations (ODEs). We furthermore apply Model Order Reduction (MOR) techniques such as Proper Orthogonal Decomposition (POD) to a network of moderate size and complexity and show that one can reduce the system size significantly.

Keywords Proper orthogonal decomposition • Nonlinear systems • Gas network simulation • Model order reduction • Differential algebraic equations

Mathematics Subject Classification (2010) 65280

1 Introduction

Natural gas, as a carrier of energy, is transported over long distances in order to be available on-site for "heating, industries and power generation" [4]. The transport task from gas suppliers to consumers is effected via networks of pipelines. We are interested in a fast and stable prediction of the dynamics of natural gas transport

S. Grundel (✉) • P. Benner
Max Planck Institute for Dynamics of Complex Technical Systems, Magdeburg, Germany
e-mail: grundel@mpi-magdeburg.mpg.de; benner@mpi-magdeburg.mpg.de

L. Jansen • C. Tischendorf
Institute of Mathematics, Humboldt Universität zu Berlin, Berlin, Germany
e-mail: lejansen@math.hu-berlin.de; caren.tischendor@math.hu-berlin.de

N. Hornung • T. Clees
Fraunhofer SCAI, Sankt Augustin, Germany
e-mail: nils.hornung@scai.fraunhofer.de; tanja.clees@scai.fraunhofer.de

S. Schöps et al. (eds.), *Progress in Differential-Algebraic Equations*, Differential-Algebraic Equations Forum, DOI 10.1007/978-3-662-44926-4_9

in pipeline networks, and therefore study the system of Differential Algebraic Equations that govern this transport problem and its numerical solution.

Our final aim is to significantly reduce the size of the system and thus save computational simulation time. In particular, we think about two application cases: On the one hand, we want to speed up the numerical solution in order to make scenario analyses possible, where the dependence of the problem on environmental or operational parameters is studied. On the other hand, we have the problem in mind to find an initial gas pressure and flow distribution within the network for a given supply and consumer demand. Such a problem can be solved if we start from a trivial solution and continuously change supply and demand during a transient simulation until the given values are met. This approach is simple but may be time-consuming and can thus benefit from a reduction in size [6].

In order to derive and improve important numerical properties of our system, we make use of the so-called Tractability Index, which essentially classifies Differential Algebraic Equations. We then apply transformation techniques to the system that increase its numerical tractability. Finally, an approach from Model Order Reduction is adapted and applied, which leads to an approximate system of reduced size.

2 Gas Transport in Pipeline Networks

Pipeline networks for gas transport can be modelled by a directed graph $\mathscr{G} = (\mathscr{A}, \mathscr{N})$ of edges $\mathscr{A}$ and nodes $\mathscr{N}$, which defines the network topology, see [4, 6]. Edges $k \in \mathscr{A}$ can e.g. represent pipes, connections, compressors, valves and regulators. Within this work we only regard a simplified network for demonstrating how Model Order Reduction can be used to significantly reduce the complexity for such a network, though, and restrict ourselves to pipe components. This simplification may be justified by several practical applications. Consider, for example, the calculation of a feasible initial solution for the computation of a more complex network. This initialization can be subdivided into a series of simpler problems (using simpler network components) that become increasingly more involved [6]. In order to speed up the initialization procedure, it can already be helpful to reduce and thus accelerate the simpler sub-problems.

The set of nodes $\mathscr{N}$ can be partitioned into supply nodes $\mathscr{N}_+$, demand nodes $\mathscr{N}_-$ and interior nodes $\mathscr{N}_0$ (sometimes called junctions)

$$\mathscr{N} = \mathscr{N}_+ + \mathscr{N}_- + \mathscr{N}_0.$$

We model gas transport within a single pipe segment by a simplification of the so-called isothermal Euler equations, see e.g. [4, 6, 7, 11, 14] for more details. The isothermal Euler equations form a coupled system of Partial Differential Equations (PDEs) consisting of mass and momentum balance laws together with a constitutive relation. In particular, our system of equations includes the continuity equation (1), the pressure loss equation (2) and the equation of the state of a real gas (3).

$$\partial_t \rho + \partial_x q = 0 \tag{1}$$

$$\partial_t q + \partial_x p + \partial_x(\rho v^2) + g\rho\,\partial_x h = -\frac{\lambda(q)}{2D}\rho v|v| \tag{2}$$

$$p = \gamma(T)\, z(p, T)\, \rho \tag{3}$$

As independent variables, we consider $t \geq 0$ (time in seconds) and $x \in [0, L]$ (space in meters), where L is the length of the segment. As dependent variables of the full isothermal equations (before simplifications), we have a density field $\rho(x,t)$ in $\mathrm{kg\,m^{-3}}$, a velocity field $v(x,t)$ in $\mathrm{m\,s^{-1}}$, a pressure field $p(x,t)$ in $\mathrm{kg\,m^{-1}s^{-2}}$ and a temperature field $T(x,t)$ in K. We also define the mass flow of the gas per unit area (flux) as $q(x,t) = \rho(x,t)\,v(x,t)$ in $\mathrm{kg\,s^{-1}m^{-2}}$ and consider a given geodesic height $h(x)$, given diameter $D(x)$, as well as a friction coefficient $\lambda(q)$, compressibility $z(p,T)$, and the field $\gamma = RT$, where R is a gas constant.

In order to simplify this system, we follow [6] and approximate γz in the dependence between p and ρ (3) by the square of the sound velocity $a \approx 300\,\mathrm{m\,s^{-1}}$ of the gas

$$\begin{aligned} \partial_t \rho + \partial_x q &= 0 \\ \partial_t q + \partial_x p + \partial_x(\rho v^2) + g\rho\,\partial_x h &= -\frac{\lambda}{2D}\rho v|v| \\ p &= a^2\,\rho. \end{aligned}$$

Additionally, we replace v by q and ρ, neglect differences in geodesic height and temperature ($\forall x : h(x) \equiv h_0, T(x) \equiv T_0$) as well es kinetic energy, and obtain

$$\begin{aligned} \partial_t \rho + \partial_x q &= 0 \\ \partial_t q + a^2 \partial_x \rho &= -\frac{\lambda}{2D}\frac{q|q|}{\rho}. \end{aligned} \tag{4}$$

Given the PDE for a pipe in this way, we need to discretize it along each individual segment. Furthermore, we consider the network graph $\mathcal{G}$ and establish a mass balance equation for all interior and demand nodes as well as the boundary conditions on the gas density for the supply nodes. These mass balance equations equate to Kirchhoff's first law in electric networks. The discretization in each pipe is achieved by simple differences, while the remaining variables are replaced by simple averages. Let $d_i(t)$ be the time dependent demand at the i-th demand node and let $s_i(t)$ be the time dependent gas density at the i-th supply node. We will denote the density at node i by ρ_i and the flow in the k-th pipe at the left by q_L^k and at the right by q_R^k. In addition, we call the density at the node to the left of the k-th pipe ρ_L^k and the density at the node to the right ρ_R^k, which leads to the following set of equations

$$0 = \rho_i(t) - s_i(t) \qquad \text{for all } i \in \mathcal{N}_+, \tag{5a}$$

$$0 = \sum_{\bar{q}_R \in I_R^i} \bar{q}_R - \sum_{\bar{q}_L \in I_L^i} \bar{q}_L \qquad \text{for all } i \in \mathcal{N}_0, \tag{5b}$$

$$0 = \sum_{\bar{q}_R \in I_R^i} \bar{q}_R - \sum_{\bar{q}_L \in I_L^i} \bar{q}_L - d_i(t) \qquad \text{for all } i \in \mathcal{N}_-, \tag{5c}$$

$$\partial_t \frac{\rho_R^k + \rho_L^k}{2} = -\frac{q_R^k - q_L^k}{L_k} \qquad \text{for all } k \in \mathcal{A}, \tag{5d}$$

$$\partial_t \frac{q_R^k + q_L^k}{2} = -a_k^2 \frac{\rho_R^k - \rho_L^k}{L_k} - \frac{\lambda_k}{4D_k} \frac{(q_R^k + q_L^k)|q_R^k + q_L^k|}{\rho_R^k + \rho_L^k} \qquad \text{for all } k \in \mathcal{A}, \tag{5e}$$

with I_R^i the set of flows at the right of the pipes which are connected at the right to the i-th node, and I_L^i the set of flows at the left of the pipes which are connected at the left to the i-th node. We furthermore denote the width and the length of the kth pipe by D_k and L_k.

We will rewrite system (5) to obtain a matrix form of it. In order to do so, we use the incidence matrix A of the directed graph for the nodes in $\mathcal{N}_- \bigcup \mathcal{N}_0$

$$(A)_{ij} \equiv \begin{cases} 1 & \text{if edge } j \text{ enters node } i, \\ -1 & \text{if edge } j \text{ leaves node } i, \\ 0 & \text{else.} \end{cases}$$

The matrix A can be interpreted as the incidence matrix of the complete network reduced by all rows related to the supply nodes. Just like any reduced incidence matrix of a graph, A has full row rank [5]. The incidence matrix has a left kernel that is of dimension 1 or less. If we, however, remove at least one row, the resulting matrix A has a trivial left kernel. Notice the analogy to electric networks. The incidence matrix of an electric network is always reduced by one reference node, which is called the mass node, see [15]. The electric potential at the mass node is fixed analogously to the gas density at a supply node.

So as to set up the matrix form of (5), we also need the incidence matrix of the directed graph for the supply nodes $\mathscr{N}_+$

$$(A_S)_{ij} \equiv \begin{cases} 1 & \text{if edge } j \text{ enters node } i\,, \\ -1 & \text{if edge } j \text{ leaves node } i\,, \\ 0 & \text{else.} \end{cases}$$

These are the rows of the complete incidence matrix which were not used in A. More precisely, if the complete incidence matrix is called B, we get in MATLAB® notation $A_S = B(\text{supply}, :)$ and $A = B(\text{nonsupply}, :)$ picking either the nodes that are supplies (to form A_S) or those that are not.

Next define the incidence matrices of the undirected graph $|A|$ and $|A_S|$ as the component-wise absolute values of the incidence matrices of the directed graph. Notice that $|A|$ also has full row rank. Since the flow of a pipe on the left side will generally not coincide with the flow on the right side we also define two partial incidence matrices of the directed graph for the nodes $\mathscr{N}_- \bigcup \mathscr{N}_0$

$$(A_L)_{ij} \equiv \begin{cases} -1 & \text{if edge } j \text{ leaves node } i\,, \\ 0 & \text{else,} \end{cases}$$

and

$$(A_R)_{ij} \equiv \begin{cases} 1 & \text{if edge } j \text{ enters node } i\,, \\ 0 & \text{else.} \end{cases}$$

These partial incidence matrices are tightly linked to the matrices A and $|A|$:

$$A = A_R + A_L \text{ and } |A| = A_R - A_L.$$

In addition to the previous definitions, let $\rho = \begin{pmatrix}\rho_s & \rho_d\end{pmatrix}$ with ρ_s the density at the supply nodes and ρ_d the density at the demand nodes and junctions. Moreover, define two material-dependent matrices $M_L = \text{diag}(\ldots \frac{L_k}{4} \ldots)$ and $M_a = \text{diag}(\ldots -\frac{a_k^2}{L_k} \ldots)$, as well as a function g, component-wise for every pipe, by

$$g_k\left(\frac{q_R^k + q_L^k}{2},\ \rho_R^k,\ \rho_L^k\right) = -\frac{\lambda_k}{D_k}\frac{\frac{q_R^k+q_L^k}{2}\left|\frac{q_R^k+q_L^k}{2}\right|}{\rho_R^k + \rho_L^k}.$$

By defining the vector of all right flows of all pipes $q_R = \begin{pmatrix}\ldots q_R^k \ldots\end{pmatrix}$ and all left flows of all pipes $q_L = \begin{pmatrix}\ldots q_L^k \ldots\end{pmatrix}$ as well as the vector of supplies

$s(t) = \big(\ldots\ s_i(t)\ \ldots\big)$ and demands $d(t) = \big(\ldots\ d_i(t)\ \ldots\big)$ where for nodes in $\mathcal{N}_0$ the demands are 0, we are finally able to denote (5) in matrix form.

$$0 = \rho_s - s(t) \tag{6a}$$

$$0 = A_R q_R + A_L q_L - d(t) \tag{6b}$$

$$|A_S|^T \partial_t \rho_s + |A|^T \partial_t \rho_d = -M_L^{-1} \frac{q_R - q_L}{2} \tag{6c}$$

$$\partial_t \frac{q_R + q_L}{2} = M_a (A_S^T \rho_s + A^T \rho_d) + g\left(\frac{q_R + q_L}{2},\ \rho_s,\ \rho_d\right) \tag{6d}$$

One can write in short form

$$E\dot{x} = Hx + f(x,t) \tag{7}$$

where $x = \big(\rho_s\ \rho_d\ q_R\ q_L\big)^T$, $f(x,t) = \Big(-s(t)\ -d(t)\ 0\ g\left(\frac{q_R+q_L}{2},\ \rho_s,\ \rho_d\right)\Big)^T$ and the matrices are given by

$$E = \begin{pmatrix} 0 & 0 & 0 & 0 \\ 0 & 0 & 0 & 0 \\ |A_S^T| & |A^T| & 0 & 0 \\ 0 & 0 & \frac{1}{2}I & \frac{1}{2}I \end{pmatrix}, \quad H = \begin{pmatrix} I & 0 & 0 & 0 \\ 0 & 0 & A_R & A_L \\ 0 & 0 & -\frac{1}{2}M_L^{-1} & \frac{1}{2}M_L^{-1} \\ M_a A_S^T & M_a A^T & 0 & 0 \end{pmatrix}.$$

The parameters in the system are $\lambda_k,\ a_k,\ L_k,\ D_k$. These parameters are known at least within some range of uncertainty. With regard to the applications we have in mind, it is crucial to make sure that an approximate model of reduced size can still handle small variations of those parameters. The functions $s(t)$ and $d(t)$ are considered as input functions. They are time-dependent, yet within a certain class of functions. One is interested in reduced-order models which give good results for a large class of such input functions.

3 Tractability Index

There are several different concepts for DAE classification, called index concepts. For a survey refer to [12]. We will use the Tractability Index due to März [10]. In particular, we will apply the concept of the Tractability Index to an equation given in the form (7). This equation is an Ordinary Differential Equation if E has full rank.

In the following, we introduce the matrices G_0, G_1, G_2 such that either G_1 does not have full rank and the system has index 2 if G_2 has full rank, or the system has index 1 if G_1 already has full rank. The sequence is given by

$$G_0 = E, \tag{8a}$$

$$G_1 = G_0 + B_0 Q_0, \tag{8b}$$

$$G_2 = G_1 + B_1 Q_1. \tag{8c}$$

Here the Q_i are projectors onto the kernel of G_i. Therefore, we see immediately that $Q_1 = 0$ and $G_2 = G_1$ if G_1 has full rank. Furthermore, $B_0 = H + J_x f$ where $J_x f$ is the Jacobian of f with respect to x, and $B_1 = B_0 P_0$ where $P_0 = 1 - Q_0$. This series will in theory also give a recipe on how to compute projection matrices to decouple the algebraic from the differential part. This is, however, computationally expensive. In the next section we will see that in the case of gas transport networks we can reduce the equations to an Ordinary Differential Equation without computing the projector matrices explicitly. The resulting ODE is easy to simulate and reduce.

4 Index Analysis and Reduction

In this section the Tractability Index of the DAE (6) will be analyzed. Furthermore, a reduction procedure will be presented such that (6) can be reduced to a system of Ordinary Differential Equations.

For the index analysis we use the invariance of the Tractability Index with respect to constant transformations and refactorizations, see [10]. We prove that the Tractability Index of (6) is equal or lower than two in two steps. In the first step, we prove that a certain DAE prototype is always of index 2 or lower, and secondly, we convert (6) into a DAE of the prototype form (9) by constant transformations and refactorizations only. Consider the following class of DAEs

$$\partial_t x_1 = f(x_1, x_2, y_1, y_2, t) \tag{9a}$$

$$0 = x_2 - s(t) \tag{9b}$$

$$0 = y_1 + g(x_1, x_2, y_2, t) \tag{9c}$$

$$M \partial_t x_2 = y_2 \tag{9d}$$

with the partial derivatives f_{x_1}, f_{x_2}, f_{y_1} and f_{y_2} of f and the partial derivatives g_{x_1}, g_{x_2} and g_{y_2} of g.

Lemma 1 *Equations of type* (9) *are always of Tractability Index 2 or lower.*

Proof The first sequence of the matrix chain of the Tractability Index is given by

$$G_0 = \begin{pmatrix} I & 0 & 0 & 0 \\ 0 & 0 & 0 & 0 \\ 0 & 0 & 0 & 0 \\ 0 & M & 0 & 0 \end{pmatrix}, \quad Q_0 = \begin{pmatrix} 0 & 0 & 0 & 0 \\ 0 & Q & 0 & 0 \\ 0 & 0 & I & 0 \\ 0 & 0 & 0 & I \end{pmatrix}, \quad P_0 = \begin{pmatrix} I & 0 & 0 & 0 \\ 0 & P & 0 & 0 \\ 0 & 0 & 0 & 0 \\ 0 & 0 & 0 & 0 \end{pmatrix},$$

$$B_0 = \begin{pmatrix} f_{x_1} & f_{x_2} & f_{y_1} & f_{y_2} \\ 0 & I & 0 & 0 \\ g_{x_1} & g_{x_2} & I & g_{y_2} \\ 0 & 0 & 0 & I \end{pmatrix}$$

with Q a projector onto $\ker M$ and $P = I - Q$ the complementary projector. Clearly we have then that $PQ = QP = MQ = 0$ and $MP = M$. Hence we obtain

$$G_1 = G_0 + B_0 Q_0 = \begin{pmatrix} I & f_{x_2}Q & f_{y_1} & f_{y_2} \\ 0 & Q & 0 & 0 \\ 0 & g_{x_2}Q & I & g_{y_2} \\ 0 & M & 0 & I \end{pmatrix}, \quad B_1 = B_0 P_0 = \begin{pmatrix} f_{x_1} & f_{x_2}P & 0 & 0 \\ 0 & P & 0 & 0 \\ g_{x_1} & g_{x_2}P & 0 & 0 \\ 0 & 0 & 0 & 0 \end{pmatrix}.$$

At this point we already see that the DAE (9) has index 1 if and only if $M = 0$. Otherwise we calculate

$$Q_1 = \begin{pmatrix} 0 & K & 0 & 0 \\ 0 & P & 0 & 0 \\ 0 & g_{y_2}M & 0 & 0 \\ 0 & -M & 0 & 0 \end{pmatrix}, \quad G_2 = G_1 + B_1 Q_1 = \begin{pmatrix} I & f_{x_1}K + f_{x_2} & f_{y_1} & f_{y_2} \\ 0 & I & 0 & 0 \\ 0 & g_{x_1}K + g_{x_2} & I & g_{y_2} \\ 0 & M & 0 & I \end{pmatrix}.$$

with $K \equiv (f_{y_2} - f_{y_1} g_{y_2})M$. Notice that G_2 is always non-singular, hence the DAE is of index 2 or lower. □

Finally, we need to show that (6) can be converted into (9) under constant transformations and refactorizations.

Lemma 2 *(i) System* (6) *has Tractability Index 2 or lower. (ii) More precisely, there exists a conversion of* (6) *by index-invariant transformations to the system*

$$\begin{aligned} \partial_t \rho_m &= -(V_1^T |A^T|)^{-1} q_1 \\ \partial_t q_+ &= \tilde{g}(q_+, \rho_s, \rho_m) \\ 0 &= \rho_s - s(t) \\ 0 &= q_1 - (|A| M_L V_1)^{-1} (d(t) - A q_+ - |A| M_L W_1 V_2 q_2) \\ 0 &= q_3 \\ -V_2^T W_1^T |A_S^T| \partial_t \rho_s &= q_2 \end{aligned} \tag{10}$$

for variables $\rho_m, q_1, q_2, q_3, q_+$ and projection matrices V_1, V_2, W_1. This system, with $M = -V_2^T W_1^T |A_S^T|$ and

$$x_1 = \begin{pmatrix} \rho_m \\ q_+ \end{pmatrix}, \quad x_2 = \begin{pmatrix} \rho_s \end{pmatrix}, \quad y_1 = \begin{pmatrix} q_1 \\ q_3 \end{pmatrix}, \quad y_2 = \begin{pmatrix} q_2 \end{pmatrix},$$

is then in the form of (9).

Proof Clearly, *(i)* follows from *(ii)*, so that it suffices to show the latter. First define

$$q_+ \equiv \frac{q_R + q_L}{2} \text{ and } q_- \equiv \frac{q_R - q_L}{2},$$

which leads to a constant transformation

$$\begin{pmatrix} q_R \\ q_L \end{pmatrix} = \begin{pmatrix} I & I \\ I & -I \end{pmatrix} \begin{pmatrix} q_+ \\ q_- \end{pmatrix}$$

such that (6) can be written as

$$\begin{aligned} |A_S^T|\partial_t \rho_s + |A^T|\partial_t \rho_d &= -M_L^{-1} q_- \\ \partial_t q_+ &= M_a(A_S^T \rho_s + A^T \rho_d) + g(q_+, \rho_s, \rho_d) \\ 0 &= A q_+ + |A| q_- - d(t) \\ 0 &= \rho_s - s(t). \end{aligned} \tag{11}$$

Next define four orthonormal bases. Let V_1 be an orthonormal basis of coker $|A| = \text{im}(|A|^T)$ and let W_1 be an orthonormal basis of ker $|A|$. Then W_1 and V_1 span the whole edge space. Let V_2 be an orthonormal basis of coker $|A_S|W_1$ and let W_2 be an orthonormal basis of ker $|A_S|W_1$. These bases could be related to the topology of the network but for our purposes this is not necessary. Notice that

$$\begin{pmatrix} V_1^T \\ V_2^T W_1^T \\ W_2^T W_1^T \end{pmatrix} = \begin{pmatrix} I & 0 \\ 0 & V_2^T \\ 0 & W_2^T \end{pmatrix} \begin{pmatrix} V_1^T \\ W_1^T \end{pmatrix} \tag{12}$$

is a constant non-singular matrix due to the definition of the orthonormal bases. Since kernel and cokernel partition the edge space into two complementary subspaces, (12) constitutes an iterated partitioning procedure. Notice also that we have $|A|W_1 = 0$ and $|A_S|W_1 W_2 = 0$. Hence we obtain

$$\begin{aligned} V_1^T |A_S^T|\partial_t \rho_s + V_1^T |A^T|\partial_t \rho_d &= -V_1^T M_L^{-1} q_- \\ V_2^T W_1^T |A_S^T|\partial_t \rho_s &= -V_2^T W_1^T M_L^{-1} q_- \\ \partial_t q_+ &= M_a(A_S^T \rho_s + A^T \rho_d) + g(q_+, \rho_s, \rho_d) \end{aligned}$$

$$
\begin{aligned}
0 &= -W_2^T W_1^T M_L^{-1} q_- \\
0 &= A q_+ + |A| q_- - d(t) \\
0 &= \rho_s - s(t)
\end{aligned}
$$

by a constant refactorization, which means we have multiplied the first equation of (11) from the left by (12). After this refactorization we also use the bases for a constant transformation

$$
q_- = M_L (W_1 W_2 q_3 + W_1 V_2 q_2 + V_1 q_1),
$$

which leads to

$$
\begin{aligned}
V_1^T |A^T| \partial_t ((V_1^T |A^T|)^{-1} V_1^T |A_S^T| \rho_s + \rho_d) &= -\, q_1 \\
V_2^T W_1^T |A_S^T| \partial_t \rho_s &= -\, q_2 \\
\partial_t q_+ &= M_a (A_S^T \rho_s + A^T \rho_d) + g(q_+, \rho_s, \rho_d) \\
0 &= -\, q_3 \\
0 &= A q_+ + |A| M_L V_1 q_1 + |A| M_L W_1 V_2 q_2 \\
&\quad + |A| M_L W_1 W_2 q_3 - d(t) \\
0 &= \rho_s - s(t).
\end{aligned}
$$

The eventual transformation is defined by

$$
\rho_m = (V_1^T |A^T|)^{-1} V_1^T |A_S^T| \rho_s + \rho_d ,
$$

which leads to the desired equations if

$$
\begin{aligned}
\tilde{g}(q_+, \rho_s, \rho_m) =& M_a (A_S^T \rho_s + A^T (\rho_m - (V_1^T |A^T|)^{-1} V_1^T |A_S^T| \rho_s)) \\
&+ g(q_+, \rho_s, \rho_m - (V_1^T |A^T|)^{-1} V_1^T |A_S^T| \rho_s).
\end{aligned}
$$

We obtain

$$
\begin{aligned}
\partial_t \rho_m &= -(V_1^T |A^T|)^{-1} q_1 && \text{(13a)} \\
\partial_t q_+ &= \tilde{g}(q_+, \rho_s, \rho_m) && \text{(13b)} \\
0 &= \rho_s - s(t) && \text{(13c)} \\
0 &= q_1 - (|A| M_L V_1)^{-1} (d(t) - A q_+ - |A| M_L W_1 V_2 q_2) && \text{(13d)} \\
0 &= q_3 && \text{(13e)} \\
-V_2^T W_1^T |A_S^T| \partial_t \rho_s &= q_2. && \text{(13f)}
\end{aligned}
$$

Notice that we can eliminate q_3 in (13a) and (13d) as a constant refactorization. With $M = -V_2^T W_1^T |A_S^T|$ and

$$x_1 = \begin{pmatrix} \rho_m \\ q_+ \end{pmatrix}, \quad x_2 = \begin{pmatrix} \rho_s \end{pmatrix}, \quad y_1 = \begin{pmatrix} q_1 \\ q_3 \end{pmatrix}, \quad y_2 = \begin{pmatrix} q_2 \end{pmatrix}$$

system (13) has the structure of the DAE class (9). Hence it has index 1 if $V_2^T W_1^T |A_S^T| = 0$ or Eq. (13f) is not existent, else it has index 2. □

Remark This condition for index 1 can be expressed by the topological condition that there is only one supply node in the system. To prove this assertion, the employed orthonormal matrices would have to be chosen with a topological meaning. One can then show that if the gas network has only one supply it is always of index 1 and if it has more supply nodes it has to be of index 2. We do not prove this in detail here as it is not of practical interest since most realistic examples will have more than one supply node and, therefore, are of index 2.

For practical computations system (13) is not necessarily well-suited, since the calculation of V_1, W_1, V_2, W_2 might increase the overall numerical error, and additionally, we do not want to invert $V_1^T |A^T|$ if the network is large.

Therefore, step back to (11)

$$\begin{aligned} |A_S^T|\partial_t \rho_s + |A^T|\partial_t \rho_d &= -M_L^{-1} q_- && \text{(14a)} \\ \partial_t q_+ &= M_a(A_S^T \rho_s + A^T \rho_d) + g(q_+, \rho_s, \rho_d) && \text{(14b)} \\ 0 &= A q_+ + |A| q_- - d(t) && \text{(14c)} \\ 0 &= \rho_s - s(t) && \text{(14d)} \end{aligned}$$

and remember that M_L^{-1} is a diagonal matrix, hence it is very cheap to invert. Multiply (14a) from the left by $|A|M_L$ and replace $|A|q_-$ and ρ_s with the help of (14c) and (14d) afterwards.

$$\begin{aligned} & |A_S^T|\partial_t \rho_s + |A^T|\partial_t \rho_d = -M_L^{-1} q_- && \text{(15a)} \\ \Rightarrow \quad & |A|\, M_L\, |A_S^T|\partial_t \rho_s + |A|\, M_L\, |A^T|\partial_t \rho_d = -|A| q_- && \text{(15b)} \\ \Rightarrow \quad & |A|\, M_L\, |A^T|\partial_t \rho_d = A q_+ - d(t) - |A|\, M_L\, |A_S^T|\partial_t s(t) && \text{(15c)} \end{aligned}$$

We need the derivative of the supply input functions to obtain (15c). Hence this step includes an index reduction. Since we only differentiate an input function, no complications like the drift-off phenomena are expected. We also replace ρ_s in (14b) with the help of (14d) and obtain a system in ρ_d and q_+.

$$\begin{aligned} |A|\, M_L\, |A^T|\partial_t \rho_d &= A q_+ - d(t) - |A|\, M_L\, |A_S^T|\partial_t s(t) && \text{(16a)} \\ \partial_t q_+ &= M_a A^T \rho_d + g(q_+, s(t), \rho_d) + M_a A_S^T s(t) && \text{(16b)} \end{aligned}$$

$|A|\, M_L\, |A^T|$ is positive definite and symmetric since $|A^T|$ has full column rank and M_L is a positive definite diagonal matrix, hence (16) is an implicit ODE of the form

$$\begin{aligned} M\,\partial_t x_1 &= Ax_2 + f_1(t) \\ \partial_t x_2 &= f_2(x_1, x_2, t) \end{aligned} \tag{17}$$

with M being symmetric and positive definite. This allows for a straightforward application of half-implicit solvers and MOR methods. After we have calculated ρ_d and q_+ by solving the Ordinary Differential Equation, the rest of the variables can be computed in a post processing step.

5 Model Order Reduction: POD

The gas transport equations we consider are given in the form of (7) and thus capture (16)

$$E\,\dot{x} = H\,x + f(x,t).$$

Given such an equation our goal is reduction by projection. This means we want to find a linear subspace in which the solution trajectory lies approximately. This subspace is defined by a projection matrix $W \in \mathbb{R}^{n \times \hat{n}}$ where hopefully $\hat{n} \ll n$. More precisely, we are interested in finding a solution $\hat{x}(t) \in \mathbb{R}^{\hat{n}}$ such that

$$x(t) \approx W\hat{x}(t).$$

We can then project the given equation onto that subspace by a Galerkin projection resulting in the reduced equation

$$W^T E W \dot{\hat{x}} = W^T H W \hat{x} + W^T f(W\hat{x}, t),$$

or in short form

$$\hat{E}\dot{\hat{x}} = \hat{H}\hat{x} + W^T f(W\hat{x}, t).$$

If E is a positive definite matrix then the reduced order matrix $\hat{E} = W^T E W$ will also be a positive definite matrix. If E is indefinite, possibly singular, problems are manifold. For example, a nonsingular E does not necessarily result in a nonsingular $\hat{E}$, which in general means that projection can lead to a DAE of higher index than the original one.

With the aim to create a reduced order model, i.e. to find the subspace or compute W, we have to pick a Model Order Reduction technique. We will employ a method that is established in many communities, the so-called Proper Orthogonal

Decomposition. It is relatively straightforward conceptually and can be used on nonlinear problems. However, the nonlinearity $f(W\hat{x},t)$ has to be evaluated completely, even though we reduce the system size. In the following, POD is briefly introduced for positive semi-definite matrices E. In Sect. 5.2 it is exactly explained how we use POD on the index-reduced gas ODE.

5.1 Proper Orthogonal Decomposition

We will explain the basic idea of POD-MOR first. This method is also referred to as the method of snapshots [13]. As explained above we are interested in finding a matrix W such that the solution trajectory is well approximated. If we assume that W has orthonormal columns $u_1, \ldots, u_r$ we are interested in solving the following optimization problem

$$\min \int_0^T \left\| x(t) - \sum_{i=1}^{r} \langle x(t), u_i \rangle u_i \right\|^2 \mathrm{d}t \quad \text{where} \quad \langle u_i, u_j \rangle = \delta_{ij}. \tag{18}$$

The norm can usually be any norm, but should be connected to an inner product. Since this problem is in general not solvable, we take snapshots at several time steps $t_0 \ldots, t_N$ and find the best approximating linear subspace by solving the following problem instead

$$\min_{u_1,\ldots,u_\ell} \sum_{k=1}^{N} \alpha_k \left\| x_k - \sum_{i=1}^{\ell} \langle x_k, u_i \rangle u_i \right\|^2 \quad \text{where} \quad \langle u_i, u_j \rangle = \delta_{ij}. \tag{19}$$

For $\alpha_k = 1$ and the inner product and norm being the standard Euclidean one, the solution to this problem is directly connected to the singular value decomposition of the matrix $Y = [x_1, \ldots, x_N]$. Given the singular value decomposition (SVD) of $Y = U \Sigma V^T$, the solution to (19) for the standard Euclidean inner product and Euclidean norm is obtained by the first ℓ singular vectors which are the first ℓ columns of U. Here we assume that the singular values in Σ are ordered by descending magnitude. Per definition of the SVD U and V are orthogonal matrices and Σ is a diagonal matrix.

Let u_i be the singular vectors of the snapshot matrix $Y = [x(t_1), \ldots, x(t_N)]$ as in [6]. Then for all time steps t_k

$$x(t_k) \approx \sum_{i=1}^{\ell} \langle x(t_k), u_i \rangle u_i = \sum_{i=1}^{\ell} \hat{x}_i(t_k) u_i = W\hat{x}(t_k),$$

where $W = [u_1, \dots, u_\ell]$. Though this relation does not yield for arbitrary time steps, we assume certain continuity properties of $x(t)$ and as a result

$$x(t) \approx W\hat{x}(t).$$

If we have weights α_k and the inner product is given by a weighted inner product

$$\langle x, y\rangle_E = x^T E y$$

for a symmetric positive definite matrix E we can still solve

$$\min_{u_1,\dots,u_\ell} \sum_{k=1}^{N} \alpha_k \left\| x_k - \sum_{i=1}^{\ell} \langle x_k, u_i\rangle_E u_i \right\|_E^2 \quad \text{where} \quad \langle u_i, u_j\rangle_E = \delta_{ij} \tag{20}$$

by a singular value decomposition. The following lemma is a known result, see for example [9].

Lemma 3 *The solution to* (20) *is given by*

$$u_i = YD^{1/2}\frac{\hat{\psi}_i}{\sigma_i},$$

where $\hat{\psi}_i$ *are the singular vectors and* σ_i^2 *the singular values of* $\hat{Y}^T\hat{Y} = D^{1/2}Y^T EYD^{1/2}$. *Here* $D = diag(\dots, \alpha_k, \dots)$.

Proof

$$\begin{aligned}
&\min_{u_1,\dots,u_\ell} \sum_{k=1}^{N} \alpha_k \left\| x_k - \sum_{i=1}^{\ell} \langle x_k, u_i\rangle_E u_i \right\|_E^2 \\
&= \min_{u_1,\dots,u_\ell} \sum_{k=1}^{N} \left\| \sqrt{\alpha_k} x_k - \sum_{i=1}^{\ell} \langle \sqrt{\alpha_k} x_k, u_i\rangle_E u_i \right\|_E^2 \\
&= \min_{u_1,\dots,u_\ell} \sum_{k=1}^{N} \left\| \sqrt{\alpha_k} E^{1/2} x_k - \sum_{i=1}^{\ell} \langle E^{1/2}\sqrt{\alpha_k} x_k, E^{1/2}u_i\rangle E^{1/2} u_i \right\|^2
\end{aligned}$$

As a result, the problem is equivalent to the one with $D = E$ being the identity considering $u_i \to E^{1/2}u_i$ and $x_k \to \sqrt{\alpha_k}E^{1/2}x_k$. This equivalence results in $E^{1/2}u_i = \hat{\phi}_i$ where $\hat{\phi}_i$ are the singular vectors of $\hat{Y} = E^{1/2}YD^{1/2}$. Given the singular value decomposition of $\hat{Y} = \hat{\Phi}\hat{\Sigma}\hat{\Psi}^T$ we get that

$$\begin{aligned}
\hat{\Phi}\hat{\Sigma}\hat{\Psi}^T &= E^{1/2}YD^{1/2} \\
\Rightarrow \quad \hat{\Phi} &= E^{1/2}YD^{1/2}\hat{\Psi}\hat{\Sigma}^{-1}.
\end{aligned}$$

This finally yields $u_i = YD^{1/2}\frac{\hat{\psi}_i}{\sigma_i}$ where σ_i is the i-th singular value of $\hat{Y}$ and $\hat{\psi}_i$ the corresponding left singular vector. Thus the $\hat{\psi}_i$ are singular vectors of the matrix $\hat{Y}^T\hat{Y} = \hat{\Psi}\hat{\Sigma}^2\hat{\Psi}^T$ and σ_i^2 its singular values. The advantage of the latter form is that $\hat{Y}^T\hat{Y} = D^{1/2}Y^TEYD^{1/2}$, which does not involve matrix functions of E such as the square root and the inverse. □

The projection matrix can therefore be computed by

$$W = YD^{1/2}\hat{\Psi}_r\hat{\Sigma}_r^{-1/2},$$

where $\hat{\Psi}_r$ are the first r columns of $\hat{\Psi}$ and $\hat{\Sigma}_r$ is the $r \times r$ upper block of $\hat{\Sigma}$. With this type of projection matrix we furthermore get $W^TEW = I_r$, the r-dimensional identity, and given a solution $x(t)$ of the full order system, we can create a solution to the one of reduced order by setting $\hat{x} = W^TEx(t)$.

5.2 *POD for Gas Transport Simulation*

The solution of the gas transport equations yields different time series for different values of the parameters and the input functions. One is interested in a reduced order model which gives good results for all or, more realistically, for a large class of input functions and a certain parameter range. In order to apply POD, we need full simulation data of the Ordinary Differential Equation assembled in a matrix Y. This is the snapshot matrix where the entries represent pressure at the nonsupply nodes and average fluxes for each edge in the graph. We consider them independently, creating matrices X_1 and X_2.

The reduction matrices are then created as above by

$$W_i = X_iU_i^r(\Sigma_i^r)^{-1/2} \quad \text{with} \quad X_i^TE_iX_i = U_i\Sigma_iV_i^T,$$

V_i^r and Σ_i^r from the truncated SVD. E_1 and E_2 are the diagonal blocks of the block-diagonal matrix E, which in our case is given by $E_1 = |A|M_L|A^T|$ and $E_2 = I$ as in (16). The final projection matrix is then given by

$$W = \begin{bmatrix} W_1 & 0 \\ 0 & W_2 \end{bmatrix}.$$

This is a block-diagonal projection matrix, i.e. we reduce the pressure and flux variables independently. A similar strategy is used in [8].

6 Numerical Examples

For first numerical tests, we consider two different pipe network examples, a very simple one from [6] and a more realistic one (see Fig. 3). Both are simulated and reduced in the index-reduced ODE version as discussed in Sect. 4. The Model Order Reduction is achieved via POD as described in Sect. 5. For all implementations we use MATLAB, the full ODE as well as the reduced ODE are solved with MATLAB's ODE Solver `ode23s`.

6.1 Simple Example

The simple pipe network has just one supply node and actually results in a system of Tractability Index 1. We can therefore simulate the DAE in its original form for a variety of demand and supply functions relatively easily. It is also possible to reduce the DAE directly resulting in a reduced DAE, which is shown in [6]. As discussed before, a direct reduction may possibly lead to systems of Tractability Index 2 or higher. If this is the case, dedicated DAE solvers for higher index problems should be employed. In fact, we have seen problems in simulating the reduced problem, e.g. if the supply functions used for simulation were very different from the ones used for generating the POD basis. Contrary to this, we have always experienced stability in simulations and Model Order Reductions based on (16) in combination with Sect. 5.2. A more detailed numerical comparison has not been conducted, but would certainly be interesting as more insight into practical implications of the Tractability Index can be expected. In the following, only results based on (16) are presented.

For a detailed description of the first example, see [6] and Fig. 1. Its topology consists of 17 nodes, among which there are 8 demand nodes and a single supply, simply connected by a net of 16 pipes. The lengths of the individual pipes are given as shown in Table 1, while the pipe diameters and friction coefficients remain

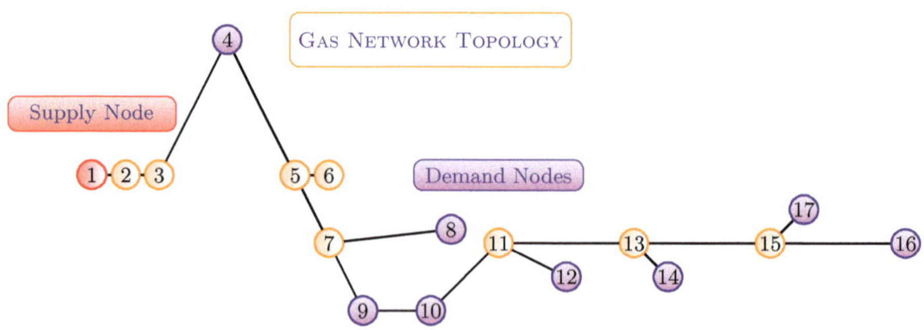

Fig. 1 Gas network graph

Table 1 Pipe lengths of the simple network, numbering in accordance with Fig. 1

Segment	$L_{1,2}$	$L_{2,3}$	$L_{3,4}$	$L_{4,5}$	$L_{5,6}$	$L_{5,7}$	$L_{7,8}$	$L_{7,9}$
Length (m)	46	7	3,080	4,318	323	790	1,820	1,460
Segment	$L_{9,10}$	$L_{10,11}$	$L_{11,12}$	$L_{11,13}$	$L_{13,14}$	$L_{13,15}$	$L_{15,17}$	$L_{15,16}$
Length (m)	2,368	1,410	296	3,979	119	3,881	687	6,114

Table 2 Fluxes at the demand nodes of the simple network, numbering in accordance with Fig. 1

Node	d_4	d_8	d_9	d_{10}	d_{12}	d_{14}	d_{16}	d_{17}
Demand ($\frac{\text{kg}}{\text{m}^2\text{s}}$)	3.15	522.97	3.3	42.46	27.15	15.6	42.76	21.75

constant throughout the entire network ($D = 0.206\,\text{m}, \lambda = 0.0003328$). The supply at time $t_0 = 0$ is given by $S_0 = 44.5\,\text{bar}$, the demands vary from node to node, see Table 2. We assume a single gas phase specified by $a_k = 430.5\,\text{m}\,\text{s}^{-1}$ for all k.

In the following scenario, we increase the demands in time from zero to the desired one in the first hour following a cosine function and then lower the supply pressure within a given time interval. Such a situation arises in practice if variations in the supply are investigated. We build the POD basis from a solution where the supply function is given by

$$s(t) = \begin{cases} S_0 & \text{if } t \leq 5{,}000\,\text{s}, \\ \left(0.95 + 0.05 \cos \frac{\pi(t-5{,}000\,\text{s})}{10{,}000\,\text{s}}\right) \times S_0 & \text{if } 5{,}000\,\text{s} < t < 10{,}000\,\text{s}, \\ 0.9 \times S_0 & \text{if } t \geq 15{,}000\,\text{s}, \end{cases} \tag{21}$$

with $t \in [0, 20{,}000\,\text{s}]$. From the data of this solution we create the POD basis as discussed. Here we take 100 time steps equally spaced in the interval from 0 to 20,000 s. We then simulate the full model and the reduced order model for a similar supply function where we lower the pressure by 20 % instead of 10 %

$$\hat{s}(t) = \begin{cases} S_0 & \text{if } t \leq 5{,}000\,\text{s}, \\ \left(0.9 + 0.1 \cos \frac{\pi(t-5{,}000\,\text{s})}{10{,}000\,\text{s}}\right) \times S_0 & \text{if } 5{,}000\,\text{s} < t < 10{,}000\,\text{s}, \\ 0.8 \times S_0 & \text{if } t \geq 15{,}000\,\text{s}. \end{cases} \tag{22}$$

This resembles – except for a small numerical error – the top curve in the pressure plot of Fig. 2a. The pressures and fluxes and their errors are plotted in Fig. 2a–d for all nodes and pipes respectively, showing neither visual difference for the fluxes nor the pressures between the original and POD reduced solution. Relative pressure errors are presented in Fig. 2b, absolute flux errors in Fig. 2d.

Since no effort has been spent on an efficient implementation of the reduced order model, it is not very useful to compare actual timings. It will be future work to use methods, e.g. DEIM [2, 3, 8], to tackle that task. By now, the size of the reduced system gives a first impression of possible savings. Therefore, Table 3 shows a quick error comparison for different reduced orders. Convergence can be observed.

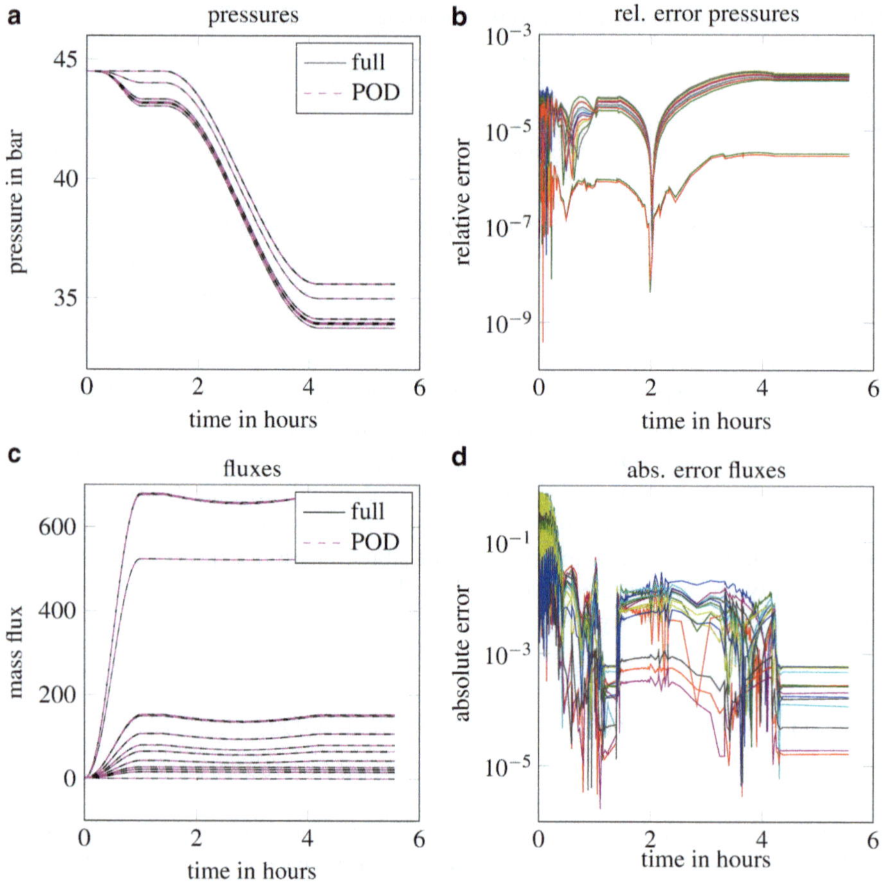

Fig. 2 Comparison of pressures and fluxes from original and POD-reduced systems of the simple network. (**a**) Comparison of pressure curves from the original and POD-reduced system for all nodes. (**b**) Pointwise relative pressure error of the POD-reduced system with respect to the original DAE solution for all nodes. (**c**) Comparison of flux curves from the original and POD-reduced system for all pipe beginnings and endings. (**d**) Pointwise absolute flux error of the POD-reduced system with respect to the original DAE solution for all pipe beginnings and endings

Table 3 Error comparison for different reduced orders, simple network

Reduction order	6	10	16
Max. rel. pressure error	1.4×10^{-4}	3.6×10^{-5}	1.5×10^{-6}
Max. abs. flux error	0.4	0.04	0.003

6.2 *Medium Size Model*

The second example, which represents a realistic part of a gas network, is a little more challenging and due to the fact that it has more than one supply it no longer has index 1. The topology of the graph is shown in Fig. 3. This system has 39 nodes and 37 pipes, which leads to an ODE system of size 72 since we have 4 supply nodes. Lengths, diameters, roughness and gas composition are chosen constant throughout ($L \equiv L_0 = 10\,\text{km}$, $D \equiv D_0 = 0.9\,\text{m}$, $k \equiv k_0 = 0.012\,\text{m}$, methane). The friction coefficient λ is approximated by the Nikuradze formula [11]. Simulating this network in the DAE form can already be slightly challenging due to its Tractability

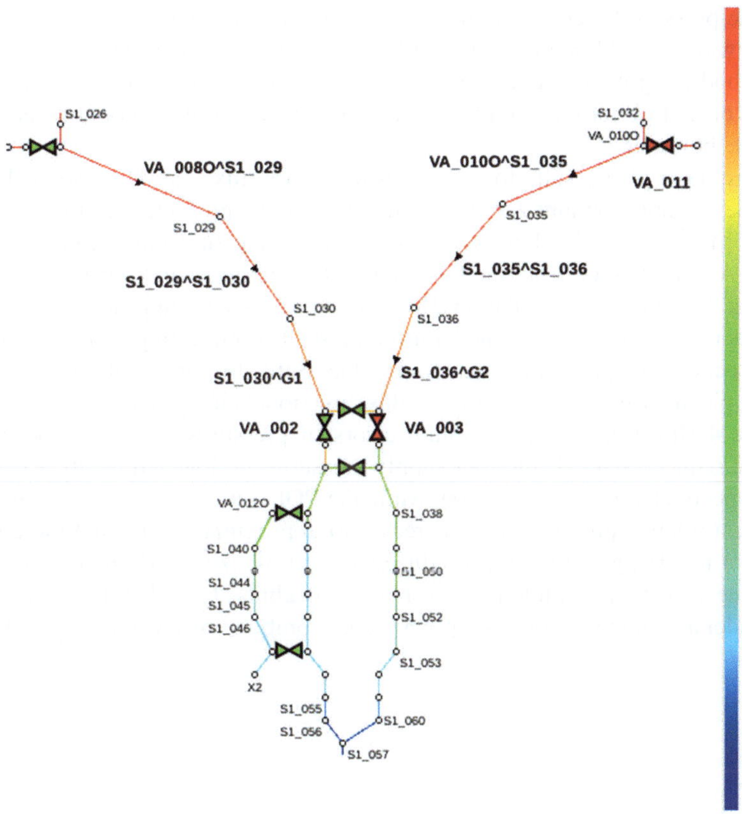

Fig. 3 Gas network graph 2

Index. We have encountered similar issues with the "naive" Model Order Reduction like for the simple model. In fact, the observed numerical issues are even more severe. In this regard, we like to point out that MOR techniques for nonlinear differential algebraic systems are not very well understood yet, and even in the case of linear problems one typically has to address the subject and take apart the differential and algebraic contributions [1].

The medium-size network of Fig. 3 has been simulated with a demand function $d(t) = d_0(1 - \cos(\pi t/3{,}600\text{s}))$ in the first hour and then constant. The network has one demand, which is set to 1,000 m^3s^{-1} (normal conditions). The supply is slowly lowered at four supply nodes as for the simple case by 10 % in order to produce snapshots for POD. We then reduce the system to size 14 and simulate again, this time reducing supplies by 20 %. All four supply nodes are described by the same pressure function. In our tests the single pipes have not been discretized further, which would be much advisable in order to achieve accurate results when the method is applied in practice. Such a discretization does not seem necessary to us in order to briefly judge the overall performance of the method by empirical evidence, however.

Figure 4a–d give results for pressures and mass fluxes at all nodes and pipes. Since the example contains zero fluxes, the flux error is presented as absolute values in $\text{kg}\,\text{s}^{-1}\text{m}^{-2}$ in Fig. 4d. For comparison, the typical flux range can be seen in Fig. 4c. As can be seen from the figures, too, the pressures and fluxes have not yet settled at the end of the simulation. However, test runs give empirical evidence that a stationary state is finally reached. Furthermore, it is interesting to observe that the flow direction changes within some pipes due to the drop in supply pressure. Flux errors typically seem to increase when this directional change happens.

Table 4 shows how the maximum errors of pressures and fluxes behave for different reduced sizes. If the four supply pressures are lowered by 20, 15, 10 and 5 %, respectively, for the evaluation, while the POD set up remains unchanged, the maximum relative pressure error increases to approximately 0.1 and the absolute flux error to 116 $\text{kg}\,\text{s}^{-1}\text{m}^{-2}$ for a reduced system of size 14. Compared to typical flux values, such an absolute flux error seems inadmissible. Therefore, we suggest to consider a certain variety of supply pressure combinations within the POD set-up already.

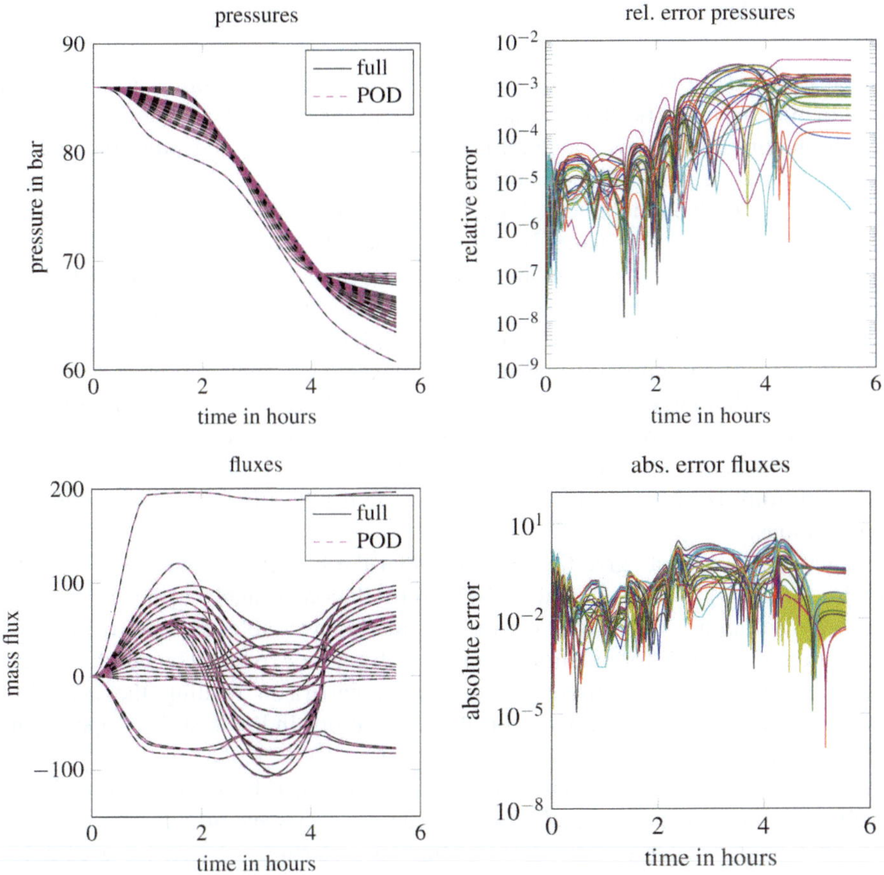

Fig. 4 Comparison of pressures and fluxes from original and POD-reduced systems of the medium network. (**a**) Comparison of pressure curves from the original and POD-reduced system for all nodes. (**b**) Pointwise relative pressure error of the POD-reduced system with respect to the original DAE solution for all nodes. (**c**) Comparison of flux curves from the original and POD-reduced system for all pipe beginnings and endings. (**d**) Pointwise absolute flux error of the POD-reduced system with respect to the original DAE solution for all pipe beginnings and endings

Table 4 Error comparison for different reduced orders, medium network

Reduction order	6	10	16	22
Max. rel. pressure error	0.0061	0.0072	0.0028	4.3836×10^{-4}
Max. abs. flux error	44.5595	8.0025	3.6276	2.3372

7 Discussion

In our work described here, we study the numerical simulation of gas transport within pipeline networks that consist of pipes only. More involved components have not been considered. Given a simplified differential system to capture the physics of gas transport, mass balance laws for the network, as well as a discretization in space, we derive a nonlinear DAE. We furthermore show that the problem lies in a class of DAEs that have Tractability Index 2 or lower, relate the index to the network topology, and give an alternative formulation with better tractability, namely an Ordinary Differential Equation. We discuss the application of Model Order Reduction by POD to the index-reduced ODE.

We also present numerical results for two examples, a smaller index 1 network and a somewhat larger index 2 network. The small network is tested in a simple setting, where it is found to have a satisfyingly accurate solution for a reduced system of size 6, when the original system size is 32. The larger network is reduced from size 72 to 14. In our opinion, these results allow for promising future applications.

Our numerical tests give first evidence that the numerical solution and Model Order Reduction of the equivalent ODE formulation is stably achieved yielding accurate solutions. We cannot report a similarly invariable success for a direct reduction of the DAE formulation – a fact which can be explained by our theoretical results. Nevertheless, a thorough empirical comparison relating the described numerical behaviour to our work on the Tractability Index is still an open issue for future research.

The simplifications we suppose are mostly motivated by industrial practice as well as by certain specific problems that can arise in scenario analysis (analysis of input or parameter variations) or in finding a starting solution for a complex network via the solution of simplified but increasingly more complex sub-problems. Nevertheless, further components ought to be considered in future research in order to allow for the Model Order Reduction of a larger class of pipeline networks from the engineering practice. The tractability of MOR systems of DAEs should perhaps be discussed and investigated more deeply in order to gain a better understanding of the matter if a system cannot be transformed as straightforwardly into an Ordinary Differential Equation as we have seen here. In future work, the topological meaning of the presented results could be considered. Also, (16) might represent an ideal starting point to discuss the effects of topology *changes* on the original system as well as on MOR approaches. We furthermore want to consider even larger networks as well as techniques such as EIM and DEIM.

References

1. Banagaaya, N., Schilders, W.: Simulation of electromagnetic descriptor models using projectors. J. Math. Ind. **3**(1) (2013). doi:doi:10.1186/2190-5983-3-1. http://www.mathematicsinindustry.com/content/3/1/1
2. Barrault, M., Maday, Y., Nguyen, N.C., Patera, A.T.: An 'empirical interpolation' method: application to efficient reduced-basis discretization of partial differential equations. C. R. Math. Acad. Sci. Paris **339**(9), 667–672 (2004). doi:10.1016/j.crma.2004.08.006
3. Chaturantabut, S., Sorensen, D.C.: Nonlinear model reduction via discrete empirical interpolation. SIAM J. Sci. Comput. **32**(5), 2737–2764 (2010). doi:10.1137/090766498
4. Ehrhardt, K., Steinbach, M.C.: Nonlinear optimization in gas networks. In: Bock, H., Phu, H., Kostina, E., Rannacher, R. (eds.) Modeling, Simulation and Optimization of Complex Processes, pp. 139–148. Springer, Berlin/New York (2005). http://www.springerlink.com/index/U8V7048500471185.pdf
5. Godsil, C., Royle, G.F.: Algebraic Graph Theory. Graduate Texts in Mathematics, vol. 207, chap. 8. Springer, New York (2001)
6. Grundel, S., Hornung, N., Klaassen, B., Benner, P., Clees, T.: Computing surrogates for gas network simulation using model order reduction. In: Koziel, S., Leifsson, L. (eds.) Surrogate-Based Modeling and Optimization, pp. 189–212. Springer, New York (2013). doi:10.1007/978-1-4614-7551-4_9
7. Herty, M., Mohring, J., Sachers, V.: A new model for gas flow in pipe networks. Math. Methods Appl. Sci. **33**(7), 845–855 (2010). doi:10.1002/mma.1197
8. Hinze, M., Kunkel, M.: Discrete empirical interpolation in POD model order reduction of drift-diffusion equations in electrical networks. In: Michielsen, B., Poirier, J.R. (eds.) Scientific Computing in Electrical Engineering SCEE 2010. Mathematics in Industry, pp. 423–431. Springer, Berlin/Heidelberg (2012). doi:10.1007/978-3-642-22453-9_45
9. Kunisch, K., Volkwein, S.: Galerkin proper orthogonal decomposition methods for a general equation in fluid dynamics. SIAM J. Numer. Anal. **40**(2), 492–515 (2002). doi:10.1137/S0036142900382612
10. Lamour, R., März, R., Tischendorf, C.: Differential Algebraic Equations: A Projector Based Analysis. Differential-Algebraic Equations Forum. Springer, Berlin/New York (2013)
11. LIWACOM Informationstechnik GmbH, Simone Research Group, Simone Software: Gleichungen und Methoden, Essen (2004)
12. Mehrmann, V.: Index concepts for differential-algebraic equations, TU, Berlin (2013, preprint)
13. Sirovich, L.: Turbulence and the dynamics of coherent structures, parts I–III. Q. Appl. Math. **45**(3), 561–590 (1987)
14. Steinbach, M.C.: On PDE solution in transient optimization of gas networks. J. Comput. Appl. Math. **203**(2), 345–361 (2007). doi:10.1016/j.cam.2006.04.018
15. Tischendorf, C.: Topological index calculation of DAEs in circuit simulation. Surv. Math. Ind. **8**(3–4), 187–199 (1999)

Index

S. Schöps et al. (eds.), *Progress in Differential-Algebraic Equations*, Differential-Algebraic Equations Forum, DOI 10.1007/978-3-662-44926-4

MIX
Papier aus verantwortungsvollen Quellen
Paper from responsible sources
FSC® C105338

If you have any concerns about our products,
you can contact us on
ProductSafety@springernature.com

In case Publisher is established outside the EU,
the EU authorized representative is:
Springer Nature Customer Service Center GmbH
Europaplatz 3, 69115 Heidelberg, Germany

Printed by Libri Plureos GmbH
in Hamburg, Germany